The Mathematics of Various Entertaining Subjects

THE MATHEMATICS OF VARIOUS ENTERTAINING SUBJECTS

RESEARCH IN RECREATIONAL MATH

◇◇

EDITED BY

Jennifer Beineke & Jason Rosenhouse

WITH A FOREWORD BY RAYMOND SMULLYAN

National Museum of Mathematics, *New York* • Princeton University Press, *Princeton & Oxford*

In the United Kingdom: Princeton University Press, 6 Oxford Street,
Woodstock, Oxfordshire OX20 1TW

press.princeton.edu

In association with the National Museum of Mathematics,
11 East 26th Street, New York, New York 10010

◈ TWO SIGMA

Jacket art: Clockwise: *Row 1*, Fig. 1: Courtesy of Anany Levitin;
Fig. 2: Cards and game, c. 1988, 1991. Cannei, LLC.
All rights reserved. SET® and all associated logos and tag lines are
registered trademarks of Cannei, LLC. Used with permission from
SET Enterprises, Inc.; Fig. 3: Courtesy of John K. McSweeney.
Row 2, Fig. 4: Courtesy of Robert Bosch; Fig. 5: Courtesy of Burr Tools/
Andreas Röver; Fig. 6: Courtesy of Anany Levitin.
Row 3, Fig. 7: Courtesy of O. Encke, www.encke.net; Fig. 8: Courtesy of
Anany Levitin; Fig. 9: Courtesy of Jennifer Beineke and Lowell Beineke.

All Rights Reserved

ISBN 978-0-691-16403-8

Library of Congress Control Number: 2015945218

British Library Cataloging-in-Publication Data is available

This book has been composed in Minion Pro

Printed on acid-free paper. ∞

Typeset by S R Nova Pvt Ltd, Bangalore, India
Printed in the United States of America

3 5 7 9 10 8 6 4

Contents

Foreword

Raymond Smullyan

Recreational mathematics can provide a most pleasant introduction to deep results in mathematics and logic to the general public. For example, take the field known as *propositional logic*, which deals with the way simple propositions are combined to form more complex propositions, using the connectives *not, and, or* (where "or" should be taken in the sense of "at least one,") *if-then*, and *if and only if*. I have found that a most useful way of approaching this subject is by the logic of *lying* and *truth-telling*, which people can relate to quite easily. And so let us turn to a place I have dubbed the Island of Knights and Knaves, in which knights always tell the truth and knaves always lie, and each native of the island is either a knight or a knave. Suppose you visit the island and come across two natives, Alan and Bob, and Alan says, about himself and Bob, "Both of us are knaves." Is Alan a knight or a knave, and what is Bob?

This problem is pretty easy: No knight could falsely claim that he and someone else were both knaves, hence Alan is obviously a knave. Since his statement is false, it is not true that both are knaves, and so at least one is a knight, and since it is not Alan, it must be Bob.

For a slightly harder problem, suppose that instead of saying, "Both of us are knaves," Alan had said, "At least one of us is a knave," (or, what amounts to the same thing, he could have said, "Either I'm a knave or Bob is a knave.) Now, what are Alan and Bob? Would you care to try solving this before reading further?

Here is the answer: If Alan were a knave, then it would be true that at least one of them was a knave, but knaves don't make true statements. Hence Alan cannot be a knave; he must be a knight. It further follows that as Alan truthfully said at least one is a knave, but since it is not Alan, it must be Bob. Hence the answer is the opposite to that of the first problem—now Alan is a knight and Bob is a knave.

The above two problems were respectively concerned with *and* and *or*. Now for one concerned with *if-then*.

Suppose Alan said, "If I am a knight, then so is Bob." Can the types of Alan and Bob now be established? Try this one!

Yes, both types can be established, and the argument is delightfully subtle! Suppose Alan is a knight. Then, as he truthfully says, if Alan is a knight, so is Bob, and since Alan *is* a knight (by supposition), Bob is also a knight.

This proves, not that Alan is a knight, nor that Bob is a knight, but only that *if* Alan is a knight, so is Bob. Thus we now know that if Alan is a knight, so is Bob, but since Alan said just that, he spoke truthfully, hence he is a knight. Further it is also true that if he is a knight, so is Bob (as he truthfully said), and since he is a knight, so is Bob. Thus, the answer is that they are both knights.

Finally, consider the case in which Alan says: "Bob and I are of the same type—both knights or both knaves." Can Alan's type be determined? What about Bob's?

We will see that the answer is that Alan's type cannot be determined, but Bob's type can be. To begin with, let us note that no native of this island can claim to be a knave, because no knight would falsely do so, and no knave would truthfully do so. Therefore, if Bob were a knave, then Alan could never have claimed to be the same type as Bob, as this would be tantamount to claiming to be a knave! Thus Bob must be a knight. As to Alan, he could be a knight, who truthfully claimed to be like Bob, or a knave, who falsely claims to be like Bob, and there is no way of telling which.

Next, I wish to consider a recreational problem that is related to a fundamental discovery in mathematical logic, which I will later tell you about.

A certain logician, whom we will call Larry, is completely accurate in his proofs—everything he can prove is really true. He visits the Island of Knights and Knaves and comes across a native named Jal, who says to him, "You can never prove that I'm a knight." Is Jal a knight or a knave? Yes, this can really be determined! Want to try?

Here is the answer, which may be a bit startling: Jal must be a knight, but Larry can never prove that he is! Here is why: If Jal was a knave, then contrary to what he falsely said, Larry *could* prove that Jal is a knight, which goes contrary to the given condition that Larry never proves false propositions. Hence, Jal cannot be a knave; he must be a knight. Hence, as he truthfully said, Larry can never prove that Jal is a knight. Thus Jal is really a knight, but Larry can never prove that he is.

This problem was suggested to me by the discovery of the great logician Kurt Gödel, that for the most powerful mathematical systems yet known, there must always be sentences which, though true, cannot be proved in the system. For each of these systems, as well as varieties of related systems, Gödel showed how to construct a sentence—call it "*G*"—which asserted its own nonprovability in the system. That is, *G* is either true but not provable in the system, or *G* is false but provable in the system. The latter alternative is ruled out by the reasonable assumption that the mathematical systems under consideration are *correct*, in the sense that only true sentences are provable in them. Thus the first alternative holds: *G* is true but not provable in the system.

How did Gödel construct a sentence that asserted its own nonprovability? The following problem should illustrate the essential idea.

Consider a computing machine that prints out expressions composed of the following five symbols: (,), *P*, *N*, and *D*. An expression is called *printable* if the machine can print it. We will call an expression X *regular* if it is compounded from just the three symbols, P, N, and D (thus it has no parentheses). For any regular expression X, by the *diagonalization* of X is meant the expression $X(X)$—for example, the diagonalization of *PNPD* is *PNPD(PNPD)*. By a *sentence* is meant any expression of one of the following forms, where X stands for any *regular* expression whatever:

1. $P(X)$
2. $NP(X)$
3. $PD(X)$
4. $NPD(X)$

These sentences are interpreted as follows (the symbols P, N, D, respectively, stand for *printable*, *not*, and *diagonalization of*):

1. $P(X)$ means X is printable, and is accordingly called *true* if (and only if) X is printable.
2. $NP(X)$ is called *true* if and only if X is not printable.
3. $PD(X)$ is called *true* if and only if the diagonalization of X is printable. (It is read, "Printable the diagonalization of X," or, in better English, "The diagonalization of X is printable.")
4. $NPD(X)$ is called *True* if and only if the diagonalization of X not printable.

We are given that the machine is totally accurate in the sense that all printable sentences are true. For example, if $P(X)$ is printable, then X is also printable, as $P(X)$ truthfully says. But suppose that X is printable. Does it follow that $P(X)$ is printable? Not necessarily! If X is printable, then $P(X)$ is true, but we are not given that all true sentences are printable; only that printable sentences are true. As a matter of fact, there is a true sentence that is not printable, and the problem is to exhibit one. Care to try?

Well, here is one: we know that for *any* regular expression X, the sentence $NPD(X)$ is true if and only if the diagonalization of X is not printable. We take the expression NPD for X, and so $NPD(NPD)$ is true if and only if the diagonalization of NPD is not printable, but the diagonalization of NPD is the very sentence $NPD(NPD)$! Thus $NPD(NPD)$ is true if and only if it is not printable, which means it is either true and not printable, or else not true but printable. The latter alternative is ruled out by the given condition that only true sentences are printable, hence it must be true that $NPD(NPD)$ is true, but the machine cannot print it.

This sentence $NPD(NPD)$ obviously corresponds to Gödel's sentence G, which asserts its own nonprovability.

This is just one example where recreational mathematics leads you naturally into the deepest topics in mathematics. In this volume, you will enjoy many other mathematical topics introduced by recreational puzzles!

"To many persons the mention of Probability suggests little else than the notion of a set of rules, very ingenious and profound rules no doubt, with which mathematicians amuse themselves by setting and solving puzzles." This was written by the British mathematician and philosopher John Venn, in his book *The Logic of Chance*, published in 1866. Indeed, whereas probability theory is today a required part of every mathematician's education, its origins are found in games of chance and the concerns of gamblers. When Blaise Pascal and Pierre de Fermat took up the Problem of Points, a gambling-themed brainteaser suggested to them by the prominent nobleman Chevalier de Méré in the mid-seventeenth century, it is doubtful they realized they were inaugurating a major new branch of mathematics.

The history of mathematics provides a litany of comparable examples. Prior to its twentieth-century emergence as a discipline central to both combinatorics and computer science, graph theory was the subject of numerous puzzles and brainteasers. William Rowe Hamilton made a tidy profit in 1857 from marketing the "Icosian game," which, in modern parlance, asked solvers to find a Hamiltonian cycle through the vertices of a dodecahedron. The problem of finding knight's tours on chessboards, today recognized as a problem in graph theory, has a history going back to the ninth century. Nor should we forget the boost given to the subject in 1735 when the great Leonhard Euler, while pondering the possibility of traversing each of the bridges in the Prussian town of Königsberg exactly once, used graph theory to resolve the question (in the negative, as it happens).

Latin squares have been studied since antiquity for their beauty and symmetry, but nowadays they arise naturally as tools in the theory of error-correcting codes and in experimental design in statistics. Non-Euclidean geometry was once scoffed at as of purely academic interest, but today it is a central tool in physics and cosmology. Large swaths of elementary number theory arose out of games and puzzles, but today it finds numerous applications in computer science, cryptography, and physics. John Conway devised the surreal numbers in the 1970s while considering certain questions arising from the game of Go. Today they are ubiquitous in combinatorial game theory, which in turn finds applications in computer science.

Get the idea? In 1959 physicist Eugene Wigner published a now-famous essay titled "The Unreasonable Effectiveness of Mathematics in the Physical Sciences." In light of our very partial list of examples, perhaps we should ask instead about the unreasonable effectiveness of recreational mathematics.

John Venn was hardly the only one to notice the amusement that ensues from setting and solving puzzles. None other than Gottfried Leibniz, co-creator of the calculus, once remarked that "Human beings are never so ingenious as when they are inventing games." Similar thoughts were expressed by the French philosopher and statesman Joseph de Maistre, who noted that "It is one of man's curious idiosyncrasies to create difficulties solely for the pleasure of resolving them." It is mysterious why that is. Anthropologist Marcel Danesi, in his book *The Puzzle Instinct*, writes

> Why have people from time immemorial been so fascinated by seemingly trivial posers, which nonetheless require substantial time and mental effort to solve, for no apparent reward other than the simple satisfaction of solving them? Is there a *puzzle instinct* in the human species, developed and refined by the forces of natural selection for some survival function? Or is this instinctual love of puzzles the product of some metaphysical force buried deep within the psyche, impelling people to behave in ways that defy rational explanation?[1]

Answering such questions is well beyond the ambitions of this book. We are content to note that Danesi was not exaggerating when he referred to "time immemorial" in discussing this topic. Among the oldest mathematical documents to have survived to the present is the Egyptian Rhind papyrus, which is largely a collection of ancient brainteasers. The isoperimetric problem is discussed by Virgil in the *Aeneid*. It was the eighth-century theologian Alcuin of York who introduced the world to the old teaser about ferrying a wolf, a goat, and a cabbage across a river. Fibonacci's 1202 book *Liber Abaci*, written at a time when mathematical research was largely dormant, contains a collection of recreational problems. This includes the rabbit problem, which produced the famous sequence of numbers that now bears his name. This historical list of puzzles can be extended for many, many pages.

It was with this history firmly in mind that the first biannual MOVES Conference was held in New York in the summer of 2013. "MOVES" is an acronym for "Mathematics Of Various Entertaining Subjects," which is to say it was a conference devoted to recreational mathematics. More than two hundred mathematicians participated, including several high school teachers and students at both the high school and college levels. They were united by their twin convictions that research into games and puzzles routinely proves useful, but also that, as mathematician and puzzle creator Henry Dudeney put it, "A good puzzle, like virtue, is its own reward."

The conference was the brainchild of Glen Whitney and Cindy Lawrence, directors of the National Museum of Mathematics (MoMath). MoMath is the nation's only museum devoted to mathematics and its many connections to

[1] M. Danesi. *The Puzzle Instinct. The Meaning of Puzzles in Human Life*, p. ix. Indiana University Press, Bloomington, 2002.

the world around us, as well as New York's only hands-on science center. The Museum was founded with the mission of changing public perceptions of mathematics, and revealing its beauty, creativity, and open-endedness. After making its debut in 2009 with the *Math Midway*, a colorful, carnival-themed traveling exhibition of math-based "edutainment," the Museum began in earnest its mission of public outreach.

The *Math Midway* comprised the Museum's first project to demonstrate the feasibility of a hands-on center devoted to mathematics. The exhibit consisted of interactive math learning experiences, such as square-wheeled tricycles, laser-based geometry exploration, and human-sized geometric puzzles; it reached more than 750,000 visitors in its five-year national tour. Around the country, museums and science centers reported an upturn in visits and field trips, with families, educators, and professional evaluators reporting an engagement with mathematics that lingered well beyond the initial experience. A spin-off of the *Math Midway, Math Midway 2 Go*, brings six of the most popular *Midway* exhibits to an even greater variety of venues, including science festivals, schools, community centers, and libraries, and has expanded MoMath's reach to include locations around the world.

In pursuit of its mission, the Museum also began offering a variety of programs and presentations to students, teachers, and the public. Highlights included programs for school groups, teacher development seminars, participation in STEM (Science, Technology, Engineering, and Mathematics) expos, sponsorship of several new middle school math tournaments, and "math tours" of cities, including Miami, New York, and Washington, DC. The Museum's *Math Encounters* program, a monthly presentation series designed to communicate the richness of mathematics to the general public, began delighting hundreds of New Yorkers month after month with its unique and engaging brand of mathematical experiences. And the *MoMath Masters*, a series of math tournaments for adults, continues to captivate contestants and audiences alike, as some of the nation's top mathematical minds battle it out for problem-solving supremacy.

After almost four years of outreach, Whitney and Lawrence, together with seasoned exhibit designer Tim Nissen, opened MoMath. Located in the heart of Manhattan's Flatiron district, the Museum's 19,000-square-foot space on the north end of Madison Square Park is home to more than three dozen hands-on exhibits and has been the site of hundreds of innovative programs for visitors of all ages. More than doubling attendance projections since it opened, MoMath provides entry to a world where kids can get excited about mathematics, and where adults and children alike can experience the evolving, creative, aesthetic, and often surprising nature of mathematics. Fast becoming a destination in New York among the technically savvy—as well as attracting tourists from around the world—MoMath was named the Best Museum for Kids by *New York* magazine in 2013 .

After opening, the Museum continued to pursue its mission, implementing a broad variety of programs with wide appeal. *Family Fridays* focuses on the entire family enjoying math together, as dynamic program leaders bring out math through hands-on activities that parents can enjoy alongside their kids. The *Power Series* program transports the world's foremost mathematicians to MoMath, to share their excitement about the current frontiers of research and knowledge. Field trips, uniquely collaborative math tournaments, summer programs, preschool sessions, and workshops for the gifted light the spark of enthusiasm in students from far and wide. Adult-based programs, including book discussions, movie nights, storytelling sessions, nights of comedy, and evening adult-only nights, highlight the wonders of mathematics beyond a traditional family-based crowd. And *Composite*, the gallery at MoMath, provides a link between the worlds of math and art, focusing on the shared beauty, creativity, and aesthetic sensibility of these two uniquely human endeavors.

Against this backdrop, MoMath reached out to the broader math community for collaboration and cooperation. The MOVES conference was conceived at a special meeting held at the 2013 Joint Mathematics Meetings, convened by Whitney and Lawrence. Officers of each of the mathematical research institutes funded by the National Science Foundation met with the two MoMath directors to discuss ways in which the Museum could maintain strong connections with the research community. Out of the numerous possible approaches contemplated, the idea of a research conference hosted at MoMath quickly gained traction with the group. In particular, a conference on recreational mathematics fit naturally with the Museum: it remains an active area in which numerous people work; there appeared to be no existing regular conference devoted to recreational math; and most importantly, like the Museum itself, it serves as a gateway for many individuals to enter the world of mathematics beyond routine scholastic arithmetic and the mechanics of algebra. MoMath was delighted to provide a home for MOVES, a conference that reminds practicing mathematicians of the fun and delight that brought them into the field and also serves to welcome new faces into the research community.

The chapters in this book represent a small portion of the work presented at the conference. They range in difficulty from those that will be accessible to any mathematically interested layperson to those whose fine details will prove challenging to all but the most dedicated readers. We believe you will find, however, that even where the details are too technical to master, a typical undergraduate math major would have sufficient background to grasp the authors' main points.

For a gentle way in, we suggest starting with Peter Winkler's delightful essay, "Should You Be Happy?" in which an often-overlooked principle of elementary probability sheds light on some challenging brainteasers. Anany

Levitin surveys a wealth of puzzles that can be solved in one move, among them some of the most ingenious conundrums ever devised. Robert Bosch, Tim Chartier, and Michael Rowan serve up a lovely synthesis of mathematics and art, as they consider various methods by which one might devise mazes that look like famous people. Jennifer Beineke and Lowell Beineke survey a museum of their own, this one composed of interesting results in graph theory.

Perhaps you are a fan of the classics? In that case you might enjoy Tanya Khovanova's discussion of new wrinkles in the genre of parallel weighing puzzles. You know the ones we mean: you have one fake coin among a collection of real ones, and you must use a balance scale a limited number of times to find the slug. Max Alekseyev and Toby Berger find new angles to explore with regard to the old Tower of Hanoi puzzle. Did you know that results from the theory of electrical circuits are relevant here? If you prefer more geometrical fare, have a look at what Julie Beier and Carolyn Yackel have to say on the subject of flexagons, which have amused mathematicians ever since their invention in 1939. Derek Smith solves a challenging family of problems inspired by certain Burr puzzles, in which your task is to assemble familiar geometric objects from various families of oddly-shaped pieces. If all of this hard work puts you in the mood for the familiar joys of a simple crossword, you might consider John McSweeney's ingenious discussion of how to model the difficulty of such a puzzle.

Playing cards are a perennial source of interest for the recreational mathematician. Dominic Lanphier and Laura Taalman examine heartless poker, by which they mean poker played with a deck lacking one of the traditional suits. Neil Calkin and Colm Mulcahy discuss the mathematics underlying certain card moves employed by magicians. Robert Vallin is likewise inspired by card tricks, using combinatorics, number theory, and analysis to study the permutations that arise from a classic effect.

Then we come to a few papers involving games. Do you think tic-tac-toe is too easy? Try playing it on an affine plane, as Maureen Carroll and Steven Dougherty do in their contribution. David Molnar sheds light on a variety of connection games. Whereas most people see the mathematical card game SET® as a pleasant diversion, Gary Gordon and Elizabeth McMahon see a contribution to the theory of error-correcting codes.

The book concludes with a pair of chapters showcasing the famous Fibonacci numbers. Leigh Marie Braswell and Tanya Khovanova contribute a paper called "The Cookie Monster Problem." If you can read that title without being instantly intrigued, then perhaps mathematics is just not for you. Stephen Lucas, meanwhile, makes a convincing case that Fibonacci numbers are not just for counting rabbits.

An eclectic mix to say the least!

It only remains to acknowledge the many people who made the MOVES Conference, and this book, possible. Pride of place must surely go to Glen Whitney and Cindy Lawrence, whose courage, persistence and vision brought MoMath to life and made the conference possible. The entire mathematical community owes them a great debt. The conference would never have happened were it not for the generous support of Two Sigma, a New York-based technology and investment company. Laura Taalman was the head organizer of the conference. That so large a conference came off without a hitch is largely the result of her heroic efforts. We must also thank the numerous mathematicians who volunteered their time to serve as peer-reviewers. The level of exposition was raised considerably by their valuable suggestions. Finally, Vickie Kearn and her team at Princeton University Press fought hard for this project, for which she has our deepest gratitude.

Enough already! There are mathematical delicacies to savor in the pages ahead, and it is time to let you get to them.

Jennifer Beineke
Enfield, Connecticut

Jason Rosenhouse
Harrisonburg, Virginia

July 31, 2015

The Mathematics of Various Entertaining Subjects

PART I

◇◇◇◇◇◇◇◇◇◇◇◇

Vignettes

1

◇◇

SHOULD YOU BE HAPPY?

Peter Winkler

The following puzzle was tested on students from high school up to graduate level. What do you think?

> You are a rabid baseball fan and, miraculously, your team has won the pennant—thus, it gets to play in the World Series. Unfortunately, the opposition is a superior team whose probability of winning any given game against your team is 60%.
>
> Sure enough, your team loses the first game in the best-of-seven series, and you are so unhappy that you drink yourself into a stupor. When you regain consciousness, you discover that two more games have been played.
>
> You run out into the street and grab the first passer-by. "What happened in games two and three of the World Series?"
>
> "They were split," she says. "One game each."
>
> Should you be happy?

In an experiment, about half of respondents answered "Yes—if those games *hadn't* been split, your team would probably have lost them both."

The other half argued: "No—if your team keeps splitting games, they will lose the series. They have to do better."

Which argument is correct—and how do you verify the answer without a messy computation?

1 Comparing Probabilities

If "should you be happy" means anything at all, it should mean "are you better off than you were before?" In the above puzzle, the question comes down to: "Is your team's probability of winning the series better now, when it needs three of the next four games, than it was before, when it needed four out of six?"

Computing and comparing tails of binomial distributions is messy but not difficult; don't bother doing it, I'll give you the results later. The aim here is to suggest another method of attack, which is called *coupling*.

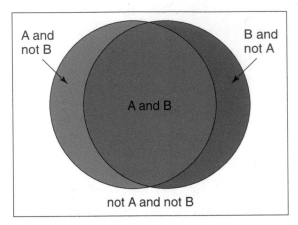

Figure 1.1. Comparing areas of the two crescents is equivalent to comparing areas of the disks.

The idea is, when you need to compare probabilities of two events A and B, to try to put them into the same experiment. All you need do is compare $\Pr(A$ but not $B)$ with $\Pr(B$ but not $A)$. This might be quite easy, especially if most of the time either both A and B occur or neither. The Venn diagram of Figure 1.1 illustrates the desired situation. If the blue region is larger than the red region, you deduce that A is more likely than B.

2 A Chess Problem, of Sorts

Let's try this on a problem adapted from Martin Gardner's legendary "Mathematical Games" column in *Scientific American*.[1] You want to join a certain chess club, but admission requires that you play three games against Ioana, the current club champion, and win two in a row.

Since it is an advantage to play the white pieces (which move first), you alternate playing white and black.

A coin is flipped, and the result is that you will be white in the first and third games, black in the second.

Should you be happy?

Gardner gave a (correct) algebraic proof of the answer ("no"), but, acknowledging the value of a proof by reasoning, he also provided two arguments that you'd be better off playing black first: (1) You must win the crucial middle

---

[1] A great source for this and lots of other thought-provoking puzzles is M. Gardner, *The Colossal Book of Short Puzzles and Problems*, W. W. Norton & Co., New York, 2006. See especially Problems 2.10 and 2.12.

game, thus you want to be playing white second; and (2) You must win a game as black, so you're better off with two chances to do so.

In fact, neither argument is convincing, and even together they are not a proof.

Using coupling, you can get the answer without algebra—even if the problem is modified so that you have to win two in a row out of *seventeen* games, or *m* in a row out of *n*. (If *n* is even, it doesn't matter who plays white first; if *n* is odd, you want to be black first when *m* is odd, white first when *m* is even.)

The coupling argument in the original two-out-of-three puzzle goes like this: Imagine that you are going to play four games against Ioana, playing white, then black, then white, then black. You still need to win two in a row, but you must decide in advance whether to discount the first game, or the last.

Obviously turning the first game into a "practice game" is equivalent to playing BWB in the original problem, and failing to count the last game is equivalent to playing WBW, so the new problem is equivalent to the old one.

But now the events are on the same space. For it to make a difference which game you discounted, the results must be either WWLX or XLWW. In words: if you win the first two games, and lose (or draw) the third, you will wish that you had discounted the last game; if you lose the second but win the last two, you will wish that you had discounted the first game.

But it is easy to see that XLWW is more likely than WWLX. The two wins in each case are one with white and one with black, so those cases balance; but the loss in XLWW is with black, more likely than the loss in WWLX with white. So you want to discount the first game (i.e., start as black in the original problem).

A slightly more challenging version of this argument works if you change the number of games played, and/or the number of wins needed in a row.

3 Back to Baseball

Let's first "do the math" and see whether you should be happy about splitting games two and three. Before the news, your team needed to win four, five, or six of the next six games. (Wait, what if fewer than seven games are played? Not to worry; we are safe in imagining that all seven games are played no matter what. It doesn't make any difference if the series is stopped when one team registers its fourth win; that, in fact, is why the rest of the games are canceled.)

The probability that your team wins exactly four of six games is "6 choose 4" (the number of ways that can happen) times $(2/5)^4$ (the probability that your team wins a particular four games) times $(3/5)^2$ (the probability that the other guys win the other two). Altogether, the probability that your team wins at least four of six is

$$\binom{6}{4}\left(\frac{2}{5}\right)^4\left(\frac{6}{5}\right)^2 + \binom{6}{5}\left(\frac{2}{5}\right)^5\left(\frac{6}{5}\right) + \binom{6}{6}\left(\frac{2}{5}\right)^6 = \frac{112}{625}.$$

After the second and third games are split, your team needs at least three of the remaining four. The probability of winning is now

$$\binom{4}{3}\left(\frac{2}{5}\right)^3\left(\frac{3}{5}\right) + \binom{4}{4}\left(\frac{2}{5}\right)^4 = \frac{112}{625}.$$

So you should be entirely indifferent to the news! The two arguments (one backward-looking, suggesting that you should be happy, the other forward-looking, suggesting that you should be unhappy) seem to have canceled each other precisely. Can this be a coincidence? Is there any way to get this result "in your head?"

Of course there is. To do the coupling a little imagination helps. Suppose that, after game three, it is discovered that an umpire who participated in games two and three had lied on his application. There is a movement to void those two games, and a countermovement to keep them. The commissioner of baseball, wise man that he is, appoints a committee to decide what to do with the results of games two and three; and in the meantime, he tells the teams to keep playing.

Of course, the commissioner hopes—and so do we puzzle-solvers—that *by the time the committee makes its decision, the question will be moot.*

Suppose five(!) more games are played before the committee is ready to report. If your team wins four or five of them, it has won the series regardless of the disposition of the second and third game results. On the other hand, if the opposition has won three or more, they have won the series regardless. The only case in which the committee can make a difference is when your team has won exactly three of those five new games.

In that case, if the results of games two and three are voided, one more game needs to be played; your team wins the series if they win that game, which happens with probability 2/5.

If, on the other hand, the committee decides to count games two and three, the series ended before the fifth game, and whoever *lost* that game is the World Series winner. Sounds good for your team, no? Oops, remember that your team won three of those five new games, so the probability that the last game is among the two they lost is again 2/5. Voilà !

4 Coin-Flipping and Dishwashing

Here's another probability-comparison puzzle, again adapted from Martin Gardner. This one yields to a variation of the coupling arguments you've seen above.

You and your spouse flip a coin to see who washes the dishes each evening. "Heads" he washes, "tails" you wash.

Tonight he tells you he is imposing a different scheme. You flip the coin thirteen times, then he flips it twelve times. If you get more heads than he does, he washes; if you get the same number of heads or fewer, you wash.

Should you be happy?

Here, the easy route is to imagine that first you and your spouse flip just twelve times each. If you get different numbers of heads, the one with fewer will be washing dishes regardless of the outcome of the next flip; so those scenarios cancel. The rest of the time, when you tie, the final flip will determine the washer. So it's still a fifty-fifty proposition, and you should be indifferent to the change in procedure (unless you dislike flipping coins).

5 Application to Squash Strategy

Squash, or "squash racquets," is a popular sport in Britain and its former colonies, probably familiar to some readers and not to others. The game is played by two players (usually) in a rectangular walled box, with slender racquets and a small black ball. As in tennis, table tennis, and racquetball, one player puts a ball in play by serving, and then the last to make a legal shot wins the rally.

In squash (using English scoring, sometimes known as "hand-out" scoring), a point is added to the rally-winner's score only when he (or she) has served. If the rally is won by the receiver, no point is scored, but the right to serve changes hands. The game is won normally by the first player to score nine points, but there is an exception: if the score reaches eight-all, the player *not* serving has the right to "set two," meaning that he may change the target score to ten points instead of nine. If he does not exercise this right, he has "set one" and the game continues to be played to nine. This choice is final; that is, even though the score may remain at eight-all for a while, no further decisions are called for.

The question (asked by Pradeep Mutalik at the recent Eleventh Gathering for Gardner, but no doubt asked by many before him) is: if you are in that position, should you set two, or just continue to play for nine points?

Simplifying the situation here (as opposed to tennis) is the fact that in squash, especially in the British ("soft ball") form of the game, having the service has almost no effect on the probability of winning the subsequent rally. Serving usually just puts the ball in play, and many strokes are likely to follow.

Thus, it is reasonable here to assume that you (the player not serving if and when eight-all is reached) have some fixed probability p of winning any given rally, regardless of who is serving. Further, the outcome of each rally can be assumed to be independent of all other rallies, and of the outcome of your decision to set one or set two.

The intuition here is quite similar to that in the World Series problem. If $p = \frac{1}{2}$, it seems clear that you want to set two, in order to minimize the service advantage. To put it another way, you are in imminent danger of losing the next rally and thus the game if you set one; this would appear to be the dominant factor in your decision.

However, if p is small, the fact that the longer game favors your (superior) opponent comes into play, and it should be better to keep the target at nine points and try to get lucky. If these arguments are correct, there ought to be some threshold value p_c (like the 40% of the World Series problem) at which you are indifferent: when $p > p_c$ you should set two, and when $p < p_c$ you should set one. What is the value of p_c?

You can solve this problem in principle by computing your probability of winning (as a function of p) when you set one, and again when you set two, then comparing those two values. This takes some work; there are infinitely many ways the squash game can continue, so you'll need to sum some infinite series or solve some equations. As before, however, we can minimize the work (and perhaps gain some insight) by coupling the two scenarios and concentrating on the circumstances in which your decision makes a difference.

Accordingly, let us assume that the game is played until someone scores ten points, even though the winner will have been determined earlier if you chose to set one. If "set one" beats "set two," that is, if you would have won if you had set one but lost if you had set two, it must be that you scored the next point but ended up with nine points to your opponent's ten. Call this event S_1.

For set two to beat set one, your opponent must be the one to reach 9-8 but then yield the next two points to you; call this event S_2.

It will be useful to have another parameter. There are several choices here, all about equally good. Let f be the probability that you "flip": that is, you score the next point even though you are not currently serving. (Notice that f applies to you only; your opponent's probability of flipping will be different unless $p = \frac{1}{2}$.)

To flip you must win the next rally, then either win the second as well, or lose it and flip. Thus

$$f = p(p + (1-p)f),$$

which we could solve for f, but let's leave it in this form for now.

Event S_1 requires that you flip, then lose the next rally, then fail twice to flip: in symbols,

$$\Pr(S_1) = f(1-p)(1-f)^2.$$

Event S_2 requires that you fail to flip, then flip, then either win the next rally or lose and then flip; thus

$$\Pr(S_2) = (1 - f)f(p + (1-p)f).$$

Both events require a flip and a failure to flip, so we can divide out by $f(1-f)$ and just compare $(1-p)(1-f)$ with $p + (1-p)f$, but notice that the latter is just f/p (from our equation for f), and the former is one minus the latter. Thus $f/p = \frac{1}{2}$, from which we get $2p = 1 - p + p^2$, $p = (3 - \sqrt{5})/2 \sim$ .381966011.

Thus, you should set two unless your probability of winning a rally is less than 38%. Since most games between players as mismatched as 62% : 38% will not reach the score of eight-all, you would not be far wrong simply to make it a policy to set two. Squash players' intuitions are apparently trustworthy: in tournament play, at least, set one has been a fairly rare choice.

English scoring is being gradually replaced by "PARS" (point-a-rally scoring) in which a point is scored regardless of whether the rally was won by the server or receiver. In PARS the game is played to eleven points, but perhaps partly in acknowledgment of the strategic facts explicated above, set two is in effect made automatic by requiring the winner to win by two points.

An important advantage to PARS is that the number of rallies in a game is somewhat less variable than in English scoring, thus it is easier to schedule matches in a tournament. If you are the underdog looking to register an upset and are permitted to select the scoring system for the game, which of the two should you choose?

I'll leave that to you.

Acknowledgement

My research is supported by NSF grant DMS-1162172.

2

ONE-MOVE PUZZLES WITH MATHEMATICAL CONTENT

Anany Levitin

This chapter provides a survey of mathematical puzzles solvable in one move. In agreement with dictionary definitions, a "move" can be either an act of moving a piece or a step taken to gain an objective. As to the puzzles' mathematical content, some of the puzzles in this survey arguably satisfy any reasonable expectation a reader might have in this regard, while a few others are included either for the sake of completeness or because of the puzzles' recreational value. The survey also suggests several research projects related to the included puzzles.

There are two main reasons for this survey. First, a majority of one-move puzzles are surprising and have the same appeal as short mathematical proofs. Second, it's a challenging task, because one-move puzzles with nontrivial mathematical content are quite rare.

One-move puzzles are grouped below according to their type. The types considered are divination puzzles, weighing puzzles, rearrangement puzzles, dissection puzzles, and folding puzzles. The chapter does not include one-question logic puzzles (e.g., Knights and Knaves) or puzzles that can be solved in one move only because of the small size of the puzzle's instance (e.g., making seven payments with links of a seven-link gold chain). I've also decided against the inclusion of equation puzzles composed of matchsticks or decimal digits, because they are neither based on nontrivial mathematics nor are their solutions typically very surprising.

The chapter concludes with answers to the puzzles highlighted in the survey.

1 Divination Puzzles

We begin our survey with a few remarkable puzzles. Although each of them is solved in one move, their solutions are both surprising and based on nontrivial mathematical facts.

(1) **One Coin Move for Freedom.** A jailer offers to free two imprisoned mathematicians—we'll call them Prisoner A and Prisoner B—if they manage to win the following guessing game: The jailer sets up an 8 × 8 board with one coin on each cell, some tails up and the others tails down. While Prisoner B is absent, the jailer shows Prisoner A the cell on the board that must be guessed by Prisoner B. Prisoner A is required to turn over exactly one coin on the board before leaving the room. Then Prisoner B enters and guesses the location of the selected cell. The prisoners are allowed to plan their strategy beforehand, but there should be no communication between them after the game begins. Can the prisoners win their freedom?

The one-dimensional version of the puzzle, an 8 × 1 board, was offered at the International Mathematics Tournament of Towns in 2007. Since then, the puzzle has appeared in the form given above on several websites and, recently, in print [20, #148].

The second puzzle is based on a very different mathematical fact.

(2) **Numbers in Boxes.** A magician performs the following trick with 100 closed boxes placed on a stage: Each box has one card. The cards are distinctly numbered 1 to 100 inclusively and randomly distributed among the boxes. Before the magician does the trick, she sends out her assistant, who opens the boxes to inspect the cards in them. The assistant has an option to exchange cards in exactly two boxes or to do nothing; in either case, he'll close all the boxes and leave the stage with no communication with the magician whatsoever. What should the assistant do so that the magician will be able to find any number from 1 to 100 given to her by opening no more than 50 boxes?

This puzzle is a one-move version of a puzzle previously discussed by Peter Winkler [32, pp. 18–20]. According to Winkler, the puzzle was posed by the Danish computer scientist Peter Bro Miltersen in conjunction with his research on the complexity of certain data structures [12] and solved by Miltersen's colleague Sven Skyum.

Our third example is due to Tom Cover [6].

(3) **Larger or Smaller.** Paula and Victor play the following game. Paula writes two distinct integers on two slips of paper, one per slip, and then conceals one slip in each hand. Victor chooses one of Paula's hands to see the number on that slip. Can Victor guess whether the number he sees is larger or smaller than the other number with a probability greater than $\frac{1}{2}$?

Peter Winkler [31] discussed this and another version of the puzzle with two random numbers selected from the uniform distribution on [0, 1], which has the opposite answer to that of Cover's version.

The next divination example is a puzzle version of a magic trick from William Simon's book [27, pp. 20–25].

> **(4) Fibonacci Sum Guessing.** Ask a friend to generate the first ten terms of a Fibonacci sequence starting with two arbitrary integers a and b (unknown to you). Your task is to ask the friend about the value of one of the terms in the sequence, after which you should be able to determine the sum of all ten terms. For example, you can ask your friend to tell you the first term, or the fourth term, or any other specific term of your choosing.

At the first glance the puzzle might appear insolvable, because for, say, a two-term Fibonacci sequence it's clearly impossible to do. But the simple generation of all ten terms of the sequence and their sum leads to an immediate solution, which requires just one arithmetical operation! It might be interesting to investigate for which values of n the sum of the first n terms of an arbitrary Fibonacci sequence can be obtained from a single term.

Our last divination puzzle is a minor rewording of an item from Fred Schuh's book [26, #83].

> **(5) Ultimate Divination.** Ask a mathematical friend to think of a positive integer. Then ask the friend just one yes/no question, after which you should be able to write down the number your friend chose.

Unlike the four puzzles preceding it, a degree of trickery is present in this puzzle's solution. It is certainly based on a mathematical fact, however, and hence deserves mention in this survey.

2 Weighing Puzzles

Identifying a counterfeit among a set of identical-looking coins is one of the oldest puzzle genres. Surprisingly, a few such puzzles can be solved with a single weighing. Here is the classic example that can be found in many puzzle books, including Martin Gardner's collection of his favorite puzzles [14, #9].

> **(6) A Stack of False Coins.** There are 10 stacks of 10 identical-looking coins. All coins in one of these stacks are counterfeit, while all coins in the other stacks are genuine. Each genuine coin weighs w grams, whereas each false coin weighs $w + 1$ grams, where w is known. There is also a one-pan pointer scale that can determine the exact weight of any number of coins. Identify the stack with the false coins in one weighing.

The puzzle can be generalized in several ways. The obvious generalization extends the number of coin stacks by one. A more interesting generalization allows any number of the stacks to be fake (see, e.g., [13, pp. 23–25], where coin stacks are replaced with large bottles of pills in the problem's wording). Averbach and Chein [1, #9.11] offered a variation of the puzzle involving a scale of limited accuracy. Fomin et al. gave two other variations: one with a two-pan scale with an arrow showing the difference between weights on the pans [10, #36], and the other with only the difference between the weights of a genuine and a false coin given [10, #39].

Here is a quite different weighing puzzle from the same Russian collection [10].

> (7) **One Questionable Coin.** Of 101 coins, 50 are counterfeit. The weight of one genuine coin is an unknown integer, while all the counterfeit coins have the same weight that differs from the weight of a genuine coin by 1 gram. Peter has a two-pan pointer scale that shows the difference in weights between the objects placed in each pan. Peter chooses one coin and wants to determine in one weighing whether it's genuine or a fake. Can he do this?

The answer is "yes," but the puzzle's solutions—there are at least two different ones!—are based on mathematical facts different from those underlying the solutions to the false-stack puzzles.

Of course, if we don't limit ourselves to puzzles that have to be solved in a single weighing, there are plenty to consider, such as the famous puzzle of determining in three weighings the unique counterfeit among 12 identical-looking coins (see, e.g., Levitin and Levitin [20, Problem 142] and the comments to its solution). Some of them are far from easy, including the still unsolved problem of identifying two counterfeit coins among n coins with the minimum number of weighings on a two-pan balance.

3 Rearrangement Puzzles

These puzzles start with an arrangement of identical items, such as coins, precious stones, or small circles, and ask a solver to obtain another configuration by moving one of the items to a new location. Here are two typical examples.

> (8) **Cross Rearrangement.** Move one coin in the cross-like arrangement of six coins to get two rows of four coins each (Figure 2.1).

Although the solution to this very old puzzle is fairly well known, it's worth pointing out that it immediately follows from the Inclusion-Exclusion Principle applied to horizontal (H) and vertical (V) rows in the desired configuration:

$$|H \cap V| = |H| + |V| - |H \cup V| = 4 + 4 - 6 = 2.$$

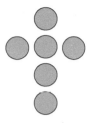

Figure 2.1. Coin arrangement for the Cross Rearrangement puzzle.

Figure 2.2. The Scholar's puzzle by Sam Loyd. © The Sam Loyd Company. The illustration is reproduced with permission of the Sam Loyd Company (samloyd.com); Sam Loyd is a registered trademark of this company.

The second example is from Sam Loyd [22, #19].

> **(9) The Scholar's Puzzle.** Erase one of the circles and draw it again at a different place on the fence to get four rows of three in a line (Figure 2.2).

The Inclusion-Exclusion Principle, though certainly not necessary, can be helpful here as well.

It's also worth mentioning here the No-Three-In-Line problem, which is thematically related to the simple puzzle above: What is the maximum number of points that can be placed in the $n \times n$ grid with no three in a straight line [30]? Although posed by Henry Dudeney almost a century ago, the final answer to the question is yet to be found.

Figure 2.3. The Giraffe puzzle.

I should also mention puzzles involving matches or toothpicks, a large subclass of rearrangement puzzles. Many one-move puzzles of this kind deal with equations typically exploiting either binary or Roman numerals but present little of mathematical content otherwise. Rather than giving specific examples of such puzzles, we conclude this section with the following remarkable brainteaser attributed to Mel Stover [15, p. 30].

> **(10) The Giraffe Puzzle.** Move one toothpick in the five-toothpick figure of a giraffe (Figure 2.3) so as to leave the giraffe exactly as it was before, except possibly for a rotation or reflection of the original figure.

4 Dissection Puzzles

These puzzles have as their objective a dissection of a given geometric figure into two pieces by a single cut. Most of them require the obtained pieces to be congruent, or have equal areas, or make it possible to reassemble the pieces into another designated figure. The cut may or may not be required to be straight; in the latter case, it may be required to go along some grid lines. Although it's unrealistic to list all the bisection puzzles included in the many puzzle collections published over the years, I'll quote a few typical examples below.

Spivak [28, #108] asks to bisect a square into (a) two congruent pentagons, (b) two congruent hexagons, and (c) two congruent heptagons. (In fact, the problem can be generalized to bisecting a parallelogram into two congruent n-gons for any $n > 2$.) For puzzles whose objective is to bisect an irregularly shaped figure into congruent halves, see [5, #47, #66], [14, #62], [21, #102], and [28, #6, #112]. Gardner [13, p. 177] gives several examples of rep-tiles—figures that can be bisected into two congruent halves that have the same shape as the original figure. Fujimura [11, #48] asks to find all the different ways to bisect a 4×4 square into two congruent pieces with cuts along small square boundaries. Hess [17, #111] offers a closely related problem of

Figure 2.4. The Yin and Yang symbol.

maximally covering polyomino shapes with two congruent tiles, the tile shape to be determined by the problem solver.

Among puzzles whose objective is to bisect a given figure into two equal-area pieces, the most mathematically interesting is the following [24, #30]:

> **(11) Triangle Bisection.** Bisect an equilateral triangle into two equal-area parts with the shortest possible dividing line.

The surprising answer is that it's not a straight line! The following remark by Polya [24, p. 272] gives the answer away: "The shortest bisector of any region is either a straight line or an arc of a circle. If the region has a center of symmetry (as the square, the circle, and the ellipse have, but not the equilateral triangle), the shortest bisector is a straight line." (An excellent research project would be to find analogous properties for area trisectors.)

Here is an area bisector example exploiting symmetries of Figure 2.4.

> **(12) Yin and Yang.** Bisect each of the black and white areas of the symbol in Figure 2.4 with one straight line [8, #158], [16, #5.14].

Since the last example involves dissecting two regions simultaneously, it's worth mentioning here the so-called *Pancake Theorem:* For any two arbitrarily shaped pancakes in the plane, there exists a single straight-line cut that divides the area of each of them exactly in half. (Of course, this puzzle-sounding theorem is just an existence result, whereas most puzzles ask for a specific cut.) The three-dimensional analog of the Pancake Theorem is known as the *Ham Sandwich Theorem:* Any sandwich made of three components of any shape—a chunk of ham and two chunks of bread, for example—can be simultaneously bisected with a single straight cut. That is, each of the three parts is cut exactly in half. Both theorems are special cases of the Stone-Tukey theorem [29]—a much more general result dealing with n measurable sets in n-dimensional space.

The following sample gives a good idea regarding puzzles about bisecting a figure into two pieces that can be reassembled into another designated figure: an $n^2 \times (n + 1)^2$ rectangle into a square [7, pp. 320–321]; mutilated rectangles into squares [7, #341], [22, #26], [8, #154], [34, #18]; a square with a removed quarter into a diamond shape [34, #5].

An easy puzzle about bisecting a chessboard asks for the maximum number of the board's squares whose interiors can be intersected by a straight-line cut [18, p. 155]. It has an obvious generalization to any $m \times n$ board, of course.

There are also a few bisection puzzles for three-dimensional shapes. For example, in 1887 Richard A. Proctor published the problem "to show how to cut a regular tetrahedron (equilateral triangular pyramid) so that the face cut shall be a square" [2, pp. 157–158]. Fujimura [11, #76] asked how to cut a cube into two congruent pieces with hexagonal cross-sections (see also [14, #42]). William Wu's online collection of puzzles [33] and the book by Savchev and Andreescu [25, p. 190] both contain the following interesting puzzle.

> **(13) Rotten Apple.** An apple is in the shape of a ball of radius 31 mm. A worm got into the apple and dug a tunnel of total length 61 mm that starts and ends at the apple's surface. (The tunnel need not be a straight line.) Prove that one can cut the apple into two congruent pieces, one of which is not wormy.

Under the name *Brick Piercing*, Wu also asks to prove the existence of a straight line that pierces the interior of a brick-filled cube, but which does not pierce any brick's interior. That is, the line only passes between the faces of adjacent bricks. (For the similar theorem for domino-tiled rectangles, proved by Ronald Graham, see [23, pp. 17–21].)

Our last example is from one of the Russian puzzle collections for mathematical circles [19, #21].

> **(14) Weird Cake.** Can one bake a cake that can be divided into four parts by a single straight cut?

5 Folding Puzzles

Folding puzzles are close relatives of dissection puzzles, although it's much harder to find meaningful one-fold puzzles than bisection puzzles. Here are two one-fold examples. The first is a simple puzzle dealing with basic shapes of plane geometry [9, #9.2].

> **(15) Three-Way Folding.** Cut from a piece of paper a shape that can be folded once to get any of the following: (a) a square; (b) an isosceles triangle; (c) a parallelogram.

The second example, tweeted recently by James Tanton, is more challenging [4].

(16) Square Folding. Fold a square piece of paper on a line that passes through the center of the square. Which line produces the least amount of overlapping area?

6 Conclusion

One-move puzzles constitute a tiny fraction of the mathematical puzzle universe. Still, it would be impossible to include all of them in this chapter, which provides a fair overview of this peculiar genre. A few of them are certainly among the best mathematical puzzles ever invented, both because of the unexpectedness of their solutions and the underlying mathematical facts.

7 Solutions

(1) One Coin Move for Freedom. The prisoners have to devise a method that will allow Prisoner A to turn one coin that will signal to Prisoner B the location of the cell selected by the jailer. To do this, they can take advantage of the locations of the coins turned, say, tails-up. More specifically, they should find a function that maps the locations of all tails-up coins into the location of the selected cell. Prisoner A's task is to turn one coin to ensure the mapping; Prisoner B's task is simply to compute the value of the function for the board presented to her. Here is how it can be done.

First, number the board's cells from 0 to 63 (e.g., going left to right along every row top to bottom). Let $T_1, T_2, \ldots, T_n$ be the six-bit binary representations of the sequential numbers assigned to all cells with tails-up coins on the board presented to Prisoner A; let J be the six-bit binary representation of the number assigned to the cell selected by the jailer. Let X be the six-bit binary representation of the sequential number assigned to the coin to be turned over by Prisoner A. To find X, find the "exclusive or" (XOR, denoted by $\oplus$) complement of the sum $T = T_1 \oplus T_2 \oplus \cdots \oplus T_n$ to J:

$$T \oplus X = J \quad \text{or} \quad X = T \oplus J. \tag{1}$$

(If $n = 0$, assume that $T = O$, the all-zero bit string, and hence $X = O \oplus J = J$.)

As an example, consider an instance of the puzzle shown in Figure 2.5. Tails-up coins and heads-up coins in the initial configuration are shown there as black and white circles, respectively; the cell selected by the jailer to be guessed is indicated by a cross. The coin to be turned over by Prisoner A, whose location is computed below, is shown by an extra circle around that coin.

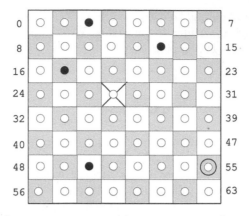

Figure 2.5. Solution to an instance of the One Coin Move for Freedom puzzle.

For this board,

$$T_1 = 2_{10} = 000010$$
$$T_2 = 13_{10} = 001101$$
$$T_3 = 17_{10} = 010001 \qquad J = 27_{10} = 011011$$
$$T_4 = 50_{10} = 110010$$

$$T = T_1 \oplus T_2 \oplus T_3 \oplus T_4 = 101100$$

and hence

$$X = T \oplus J = 101100 \oplus 011011 = 110111 = 55_{10}.$$

So, after Prisoner A turns the coin at location 55 tails-up, Prisoner B will see the board with the tails-up coins at locations 2, 13, 17, 50, and 55 and compute the location of the selected cell as

$$T_1 \oplus T_2 \oplus T_3 \oplus T_4 \oplus X = 101100 \oplus 110111 = 011011 = 27_{10} = J.$$

The above example demonstrates the first of the two possible cases—the case where the coin at location X computed by formula (1) is heads-up. Then turning it over indeed adds X to the other tails-up coins. But what if the coin at that location is already tails-up (i.e., it's the ith tails-up coin for some $1 \leq i \leq n$)? In this case, turning that coin over makes it heads-up, leaving Prisoner B to compute the location of the selected cell as $T_1 \oplus \cdots \oplus T_{i-1} \oplus T_{i+1} \oplus \cdots \oplus T_n$. Fortunately, formula (1) works in this case as well, because $S \oplus S = O$ for any bit string S. Indeed, if Prisoner A computes $X = T \oplus J = T_i$, Prisoner B will

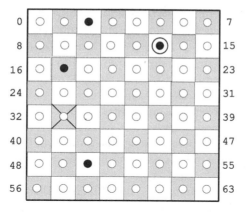

Figure 2.6. Solution to another instance of the One Coin Move for Freedom puzzle.

get the same location of the selected cell as the one used by prisoner A:

$$J = T \oplus X = T_1 \oplus \cdots \oplus T_{i-1} \oplus T_i \oplus T_{i+1} \oplus \cdots \oplus T_n \oplus T_i$$
$$= T_1 \oplus \cdots \oplus T_{i-1} \oplus T_{i+1} \oplus \cdots \oplus T_n \oplus T_i \oplus T_i$$
$$= T_1 \oplus \cdots \oplus T_{i-1} \oplus T_{i+1} \oplus \cdots \oplus T_n.$$

For example, if the jailer selects cell 33 for the board with the same four tails-up coins as shown in Figure 2.6 then we have

$$T_1 = 2_{10} = 000010$$
$$T_2 = 13_{10} = 001101$$
$$T_3 = 17_{10} = 010001 \quad J = 33_{10} = 100001$$
$$T_4 = 50_{10} = 110010$$
$$\overline{}$$
$$T = 101100$$

and hence

$$X = T \oplus J = 101100 \oplus 100001 = 001101 = T_2 = 13_{10}.$$

After Prisoner A turns the coin at cell 13 heads-up, Prisoner B will see the board with the tails-up coins at locations 2, 17, and 50, and will compute the location of the selected cell as

$$000010 \oplus 010001 \oplus 110010 = 100001 = 33_{10} = J.$$

(2) Numbers in Boxes. Number the boxes consecutively from 1 to 100, and consider the numbers placed in them as a permutation of the box numbers. This permutation can be represented in the so-called *cycle notation,* breaking the permutation into disjoint cycles. For a smaller example of numbers 1 to 8 put in boxes 3, 5, 1, 2, 8, 7, 4, 6, respectively, the two cycles are (1, 3) and (2, 4, 7, 6, 8, 5); that is, 1 is mapped into 3 and 3 into 1 in the first cycle, whereas 2 is mapped into 4, 4 into 7, 7 into 6, 6 into 8, 8 into 5, and 5 into 2 in the second cycle. Note that each cycle in the cycle notation is usually specified by a list starting with its smallest number.

To solve the problem, the assistant should identify the permutation's longest cycle. If that cycle comprises 50 elements or fewer, nothing needs to be done; otherwise, he should exchange the number in the cycle's fiftieth box with the number equal to the smallest element in the cycle in order to create a shorter cycle. (For the eight-box example above, he would exchange number 8 in the sixth box with number 2 in the fifth box.) This exchange guarantees that the longest cycle in the new permutation contains just 50 elements. The magician will then search for any given number i between 1 and 100 by starting with opening the ith box and following the mapping links indicated by the numbers in the open boxes. Since each cycle now contains no more than 50 elements, she will have to open no more than 50 boxes to reach the box with number i.

(3) **Larger or Smaller.** The surprising answer is "yes." Following Winkler's solution [31, pp. 304–305], this is how it can be done. First, Victor selects a probability distribution on the integers with positive probability assigned to each integer. After Paula writes her two numbers, Victor selects an integer from his distribution and adds $\frac{1}{2}$ to it to get a number to be denoted by t. (Note that t will not be an integer.) When Paula offers her two hands, Victor chooses one of them with probability $\frac{1}{2}$ by, say, flipping a fair coin. After inspecting the number in that hand, he guesses that the concealed number is smaller than it if the revealed number is larger than t; and he guesses that the concealed number is larger than the revealed one if the latter is smaller than t.

What can we assert about the probability that Victor's guess is correct? If t is larger than both Paula's numbers, the revealed number will be less than t, and hence Victor will guess that the concealed number is larger than the revealed one regardless of which hand he chooses. This answer will be right with probability $\frac{1}{2}$. Similarly, if t is smaller than both Paula's numbers, the revealed number will be larger than t, and hence Victor will guess that the concealed number is smaller than the revealed one regardless of which hand he chooses. This answer will be right with probability $\frac{1}{2}$ as well. But with some positive probability, Victor's t will fall between Paula's two numbers, and he will guess correctly regardless of which hand he chooses.

(4) **Fibonacci Sum Guessing.** Table 2.1 contains values of the first ten terms f_n of the sequence and their partial sums S_n. Since $S_{10} = 11 f_7$, asking for the value of f_7 allows to compute the value of S_{10} with one multiplication.

TABLE 2.1.

The first ten terms of the Fibonacci sequence starting with the numbers a and b, along with the first ten partial sums of this sequence

n	1	2	3	4	5	6	7	8	9	10
f_n	a	b	$a+b$	$a+2b$	$2a+3b$	$3a+5b$	$5a+8b$	$8a+13b$	$13a+21b$	$21a+34b$
S_n	a	$a+b$	$2a+2b$	$3a+4b$	$5a+7b$	$8a+12b$	$13a+20b$	$21a+33b$	$34a+54b$	$55a+88b$

(5) Ultimate Divination. Ask the friend if the chosen number is 1. If the answer is "yes," write "1"; otherwise, write "10," which is equal to the chosen number b in the base-b numeral system for any $b > 1$.

(6) A Stack of False Coins. Take one coin from the first stack, two coins from the second, and so on until all ten coins are taken from the last stack. Weigh all these coins together to find their excess weight over $55w$ (which is the weight of $1 + 2 + \cdots + 10 = 55$ genuine coins). The excess weight, in number of grams, indicates the number of false coins being weighed and hence the stack they were taken from.

(7) One Questionable Coin. Yes, Peter can determine whether the chosen coin is genuine. One way to do this is to put the chosen coin in one pan of the scale and all 100 other coins in the other. If the chosen coin is a fake of weight f grams while each genuine coin weighs g grams, then the difference between the weights will be

$$51g + 49f - f = 51g + 48f = 51g + 48(g \pm 1) = 99g \pm 48,$$

which is always divisible by 3. If the chosen coin is genuine, then the difference between the weights will be

$$50g + 50f - g = 49g + 50f = 49g + 50(g \pm 1) = 99g \pm 50,$$

which is not divisible by 3.

An alternative solution is given by Fomin et al. [10, pp. 217–218]. Peter puts the chosen coin aside, divides the remaining coins into two piles of 50 coins each, and weighs these piles against each other. One can show that if the chosen coin is genuine, the difference between the weights of the piles must be even, otherwise the difference must be odd.

(8) Cross Rearrangement. Double the coin at the intersection of the horizontal and vertical rows by moving there one of the other three coins in the vertical row.

(9) The Scholar's Puzzle. Erase the leftmost circle and draw a new one at the intersection point of the lines through the centers of the two highest and two lowest circles, as shown in Figure 2.7.

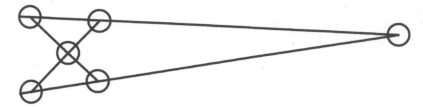

Figure 2.7. Solution to the Scholar's puzzle.

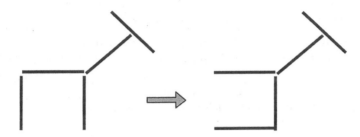

Figure 2.8. Solution to the Giraffe puzzle.

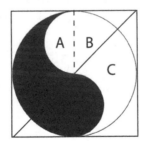

Figure 2.9. Solution to the Yin and Yang puzzle.

(10) The Giraffe Puzzle. Rotate the left leg 90° clockwise around its lower endpoint to get the giraffe shown Figure 2.8.

(11) Triangle Bisection. The shortest line bisecting an equilateral triangle into two equal-area parts is the arc of the circle with center at the triangle's vertex and radius r satisfying the equation $\frac{1}{6}\pi r^2 = \frac{\sqrt{3}}{8}a^2$, where a is the length of the triangle's side. For a proof based on the Isoperimetric Inequality, see Bogomolny [3].

(12) Yin and Yang. The diagonal of the square around the Yin-Yang symbol (Figure 2.9) bisects each the black and white areas in it.

Figure 2.10. Solution to the Weird Cake puzzle.

The following simple proof that it is the case has been sent to Martin Gardner by readers of his *Scientific American* column [16, p. 126]. The area of the small semicircle A is clearly 1/8 that of the large circle, because the radius of the former is half the size of the latter. The area of region B is also 1/8 that of the large circle's area. Thus the area of A and B together is 1/4 that of the large circle, the same as the area of region C. Hence the diagonal line does bisect both the Yin and Yang regions.

(13) Rotten Apple. Let S and F be the beginning and end points of the tunnel dug by the worm. Following Savchev and Andreescu [25, p. 191], consider the set of points X such that $|XS| + |XF| \leq 61$, where $|XS|$ and $|XF|$ are the Euclidean distances from point X to points S and F, respectively. This set is an ellipsoid of rotation with foci S and F. Each point of the tunnel belongs to this ellipsoid. Indeed, $|SX|$ and $|XF|$ don't exceed the lengths of the tunnel's parts from S to X and from X to F, respectively, and therefore $|SX| + |XF| \leq 61$. However, the center O of the apple does not belong to the ellipsoid, since $|OS| + |OF| = 31 + 31 = 62 > 61$. But the ellipsoid is convex, and therefore there exists a plane through point O that has no common points with the ellipsoid. (This geometrically obvious fact also follows from a corollary of the Hyperplane Separation Theorem that guarantees the existence of a hyperplane separating two disjoint compact convex sets in n-dimensional Euclidean space.) Such a plane will cut the apple into two congruent pieces, one of which is not wormy.

(14) Weird Cake. Among many possibilities, the cake in question can have the shape of the capital letter E (Figure 2.10).

(15) Three-Way Folding. A trapezoid composed of a square and a right isosceles triangle satisfies all three requirements of the puzzle. Foldings along the dashed lines in Figure 2.11 yield a square, an isosceles triangle, and a parallelogram, respectively.

(16) Square Folding. The line producing the least amount of overlapping area has a slope of 22.5°. Assuming the square has side length of 1, the size of the minimal overlapping area is equal to $\sqrt{2} - 1$, while $x = y = 1 - \frac{1}{\sqrt{2}}$ in Figure 2.12. For a proof, see Bogomolny [4].

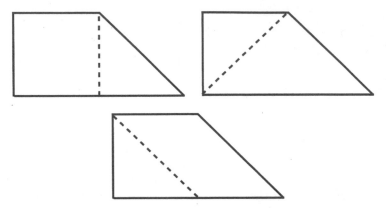

Figure 2.11. Solution to the Three-Way Folding puzzle.

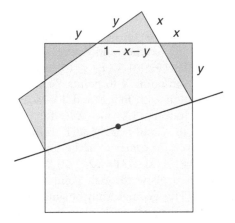

Figure 2.12. Solution to the Square Folding puzzle.

References

[1] B. Averbach and O. Chein. *Problem Solving through Recreational Mathematics.* Dover, New York, 1980.

[2] C. Birtwistle. *Mathematical Puzzles and Perplexities.* George Allen & Unwin, Crows Nest, Australia, 1971.

[3] A. Bogomolny. Bisecting arcs, in *Interactive Mathematics Miscellany and Puzzles*, http://www.cut-the-knot.org/proofs/bisect.shtml (accessed August 19, 2014).

[4] A. Bogomolny. Folding square in a line through the center, in *Interactive Mathematics Miscellany and Puzzles*, http://www.cut-the-knot.org/Curriculum/Geometry/GeoGebra/TwoSquares.shtml#solution (accessed August 19, 2014).

[5] B. Bolt. *The Amazing Mathematical Amusement Arcade*. Cambridge University Press, Cambridge, 1984.

[6] T. Cover. Pick the largest number, in T. Cover and B. Gopinath, editors, *Open Problems in Communication and Computation*, p. 152. Springer-Verlag, New York, 1987.

[7] H. Dudeney. *536 Puzzles & Curious Problems*. Charles Scribner's Sons, New York, 1967.

[8] H. Dudeney. *Amusements in Mathematics*. Dover, New York, 1970.

[9] M. A. Ekimova and G. P. Kukin. *Zadachi na Razrezanie [Cutting Problems]*, 3rd ed. Moscow Center for Continuous Mathematical Education, Moscow, 2007 (in Russian).

[10] D. Fomin, S. Genkin, and I. Itenberg. *Mathematical Circles (Russian Experience)*. American Mathematical Society, Providence, RI, 1996.

[11] K. Fujimura. *The Tokyo Puzzles*. Charles Scribner's Sons, New York, 1978.

[12] A. Gal and P. B. Miltersen. The cell probe complexity of succinct data structures, in *Proceedings of Automata, Languages, and Programming: 30th International Colloquium, ICALP 2003, Eindhoven, The Netherlands, June 30–July 4, 2003*, p. 332. Lecture Notes in Computer Science 2719. Springer, New York, 2003.

[13] M. Gardner. *aha!Insight*. Scientific American/W.H. Freeman, New York, 1978.

[14] M. Gardner. *My Best Mathematical and Logic Puzzles*. Dover, New York, 1994.

[15] M. Gardner and Mel Stover, in D. Wolfe and T. Rodgers, editors, *Puzzlers' Tribute: A Feast for the Mind*, p. 29. A. K. Peters, Natick, M.A., 2002.

[16] M. Gardner. *The Colossal Book of Short Puzzles and Problems*. W. W. Norton, New York, 2006.

[17] D. Hess. *Mental Gymnastics: Recreational Mathematics Puzzles*. Dover, Mineola, NY, 2011.

[18] B. A. Kordemsky. *Matematicheskie Zavlekalki [Mathematical Charmers]*. Onyx, Moscow, 2005 (in Russian).

[19] E. G. Kozlova. *Skazki and Podskazki [Fairy Tales and Hints: Problems for a Mathematical Circle]*, second edition. Moscow Center for Continuous Mathematical Education, Moscow, 2004 (in Russian).

[20] A. Levitin and M. Levitin. *Algorithmic Puzzles*. Oxford University Press, Oxford, 2011.

[21] S. Loyd. *Mathematical Puzzles of Sam Loyd*, selected and edited by M. Gardner. Dover, New York, 1959.

[22] S. Loyd. *More Mathematical Puzzles of Sam Loyd*, selected and edited by M. Gardner. Dover, New York, 1960.

[23] G. E. Martin. *Polyominoes: A Guide to Puzzles and Problems in Tiling*. Mathematical Association of America, Washington, DC, 1996.

[24] G. Polya. *Mathematics and Plausible Reasoning*, Volume 1. Princeton University Press, Princeton, NJ, 1954.

[25] S. Savchev and T. Andreescu. *Mathematical Miniatures*. Mathematical Association of America, Washington, DC, 2003.

[26] F. Schuh. *The Master Book of Mathematical Recreations*. Dover, New York, 1968.

[27] W. Simon. *Mathematical Magic*. Dover, Mineola, NY, 2012.

[28] A. V. Spivak. *Tysjacha i Odna Zadacha po Matematike [One Thousand and One Problems in Mathematics]*. Prosvetschenie, Moscow, 2002 (in Russian).

[29] A. H. Stone and J. W. Tukey. Generalized "sandwich" theorems. *Duke Math. J.* **9**, no. 2 (1942) 356–359.

[30] Wikipedia: The Free Encyclopedia. No-three-in-line problem (accessed August 19, 2014).

[31] P. Winkler. Games people don't play, in D. Wolfe and T. Rodgers, editors. *Puzzlers' Tribute: A Feast for the Mind*, p. 301. A K Peters, Wellesley, MA, 2002.

[32] P. Winkler. *Mathematical Mind-Benders*. A K Peters, Wellesley, MA, 2007.

[33] W. Wu. Rotten apple. *Online Collection of Puzzles.* www.ocf.berkeley.edu/~wwu/riddles/hard.shtml (accessed August 19, 2014).

[34] N. Yoshigahara. *Puzzles 101: A Puzzlemaster's Challenge*. A K Peters, Wellesley, MA, 2004.

3

◇◇◇

MINIMALIST APPROACHES TO
FIGURATIVE MAZE DESIGN

Robert Bosch, Tim Chartier, and Michael Rowan

To walk safely through the maze of human life, one needs the light of
wisdom and the guidance of virtue.

—Siddhartha Gautama, the Buddha

After a long but satisfying day of teaching mathematics, you return home to
kids who are hungry for food and entertainment. Preparing dinner will be
easy, as you have shelves full of cookbooks that describe how to make delicious
meals from a half-dozen ingredients using minimal amounts of time and effort.
And normally, entertaining would be easy too. But this morning you had
left a book [11] open to a page showing woodcuts of the sixteenth-century
Paduan architect Francesco Segala, and now your children want you to be like
Francesco and make them mazes that look like their dog, their cat, or—even
better—one of them.

You tell the kids to draw their own mazes while you make dinner. As you
chop vegetables, you think about the problem. If you had more time, you could
attempt to emulate Xu and Kaplan [16] and create a software package that
allows the maze designer to partition the source image into regions, assign
textures to the regions, and specify a solution path. The output would be a
perfect maze—one with a unique path connecting the designated starting and
ending points—that has the desired textures More importantly, when viewed
from a distance, it would closely resemble the source image.

You discard this idea. Emulating Xu and Kaplan would be akin to trying
to cook like a master chef (Ferran Adrià, José Andrés, or René Redzepi, for
example). Instead, you decide to follow the same approach you are using with
dinner: you will see if you can use relatively simple mathematical techniques
to make some aesthetically pleasing figurative mazes.

1 TSP Art Mazes

One simple method, introduced by Chartier [4], is based on TSP Art [3, 10]. This method involves converting the target image into a stipple drawing, as in Figure 3.1a, and then treating the dots as the cities of a Traveling Salesman Problem (TSP), in which a salesperson, based in one of the cities, must visit each of the other cities exactly once and then return home. It is assumed that the salesperson can and will travel as the crow flies. As a result, the salesperson's tour will be polygonal, as in Figure 3.1b, and the cost incurred by the salesperson in traveling from any city i to any other city j can be measured by the Euclidean distance from dot i to dot j. This makes the TSP a geometric TSP. It is well known that for a geometric TSP, optimal tours will be simple closed curves. In other words, they won't intersect themselves. The Jordan Curve Theorem then guarantees that the saleperson's tour will divide the plane into two regions, an inside and an outside, as in Figure 3.1c. The maze's paths will lie in the inside (white) region. The final step is to remove two segments of the tour to form the maze's entrance and exit, as in Figure 3.1d. To achieve greater contrast in Figure 3.1d, we varied the dot sizes and edge thicknesses in accordance with the brightness levels of the target image.

At the present time, there is no known polynomial-time algorithm for finding a provably optimal tour, and many mathematicians and computer scientists believe that no such algorithm exists. In fact, if one *were* to exist, then this would imply that P equals NP. Consequently, one might easily come to the erroneous conclusion that the TSP Art method is doomed to failure.

Fortunately, linear-programming (LP)-based methods [1, 5, 6] have been used to solve TSPs of world-record-setting size (tens of thousands of cities), and many hundreds of heuristics have been designed to search for high-quality tours. Often, the LP-based methods can be used to prove that a tour found by a heuristic is close to optimal.

The Concorde TSP Solver [7] is computer code that includes both a state-of-the-art implementation of an LP-based method for finding optimal tours and numerous heuristics for quickly finding tours. It is available for free on many platforms, including iPhone® and iPad®.

By using Concorde, we were able to find an optimal solution to the 2,000-city TSP shown in Figure 3.1a. The optimal tour displayed in Figures 3.1b,c has length 962,842. But Concorde took nearly a half a day of computation to find it and prove that it is optimal! When we used Concorde's Lin-Kernighan heuristic (described in Cook [5]), we quickly found (in less than 10 seconds) a tour of length 963,584, which is only 742 distance units (approximately 0.07706%) longer than the optimal tour. The Lin-Kernighan tour is displayed in Figure 3.2a.

To the serious maze designer, the most significant drawback of the TSP Art method is that the resulting mazes do not cover nearly as much space

Figure 3.1. The TSP Art method (a) starts with a stipple drawing and (b) finds a tour for the corresponding TSP. (c) The tour separates the plane into two regions. (d) By removing two edges, a maze is produced.

as they could cover. The pathways of a TSP Art maze lie within the inside of the tour, the white regions in Figures 3.1c and 3.2a. To remedy this, we can remove certain edges of the tour. The outside of the tour (the gray regions in Figures 3.1c and 3.2a) can be thought of as a gray ring together with "tendrils" of gray that wind their way into the interior. To enable a maze walker to reach the region inside one particular tendril, we need to remove at least one of the edges that form its boundary. To keep the maze perfect, we must remove precisely one of these tendril edges. (If we were to remove more, we would introduce one or more cycles.) Figure 3.2b displays a perfect maze produced through this "surgical reparation" procedure.

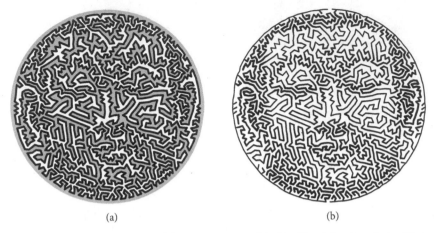

(a)   (b)

Figure 3.2. (a) A tour obtained by using Concorde's Lin-Kernighan heuristic. (b) A perfect maze produced by removing one edge from each tendril.

2 MST Art Mazes

Another approach, first presented by Inoue and Urahama [9], involves replacing the TSP with a much easier optimization problem, the problem of finding a minimum spanning tree (MST), a minimum-cost subgraph that connects all the vertices (dots) but is free of cycles. Finding an MST is easy, as there exist polynomial-time algorithms (e.g., Kruskal's algorithm and Prim's algorithm) for constructing MSTs [6].

With this approach, we have a choice: we can use the edges of the MST as the pathways of our maze, or we can use them as the walls. If we choose the former, we start with a white-on-black stipple drawing of the target image, as shown in Figure 3.3a. We then use Kruskal's algorithm to construct an MST, as shown in Figure 3.3b. To improve the contrast, we can vary the dot sizes and edge thicknesses in accordance with the brightness levels of the target image, as shown in Figure 3.3c. The final step is to attach an entrance and an exit to the tree, as in Figure 3.3d.

Our preferred variation of Inoue and Urahama's approach is to use edges as walls and use a modification of Kruskal's algorithm to construct a *one-tree*, a tree together with one cycle. We start with a black-on-white stipple drawing, as in Figure 3.4a. Then, instead of starting Kruskal's with no edges and repeatedly inserting the shortest remaining edge that does not cause a cycle to be formed, we start it with a collection of edges that join together into what could be called a "border cycle," as shown in Figure 3.4b. By repeatedly inserting the shortest remaining edge that does not cause an additional cycle to be formed, we end up constructing a minimum-cost one-tree that contains the border edges,

(a)

(b)

(c)

(d)

Figure 3.3. The MST Art method with edges as pathways: (a) a stipple drawing; (b) an MST; (c) the same MST drawn with variable dot sizes and edge thicknesses; and (d) the resulting maze.

as shown in Figure 3.4c. We then delete two border edges to form an entrance and exit, as in Figure 3.4d.

While both approaches produce beautiful images, the edges-as-walls approach produces images that have much wider and more maze-like path ways.

3 Maze Design via Phyllotaxis

To be honest, it is a stretch to label the methods described above as "min-imalist" approaches to figurative maze design. After all, to use the TSP Art

(a)

(b)

(c)

(d)

Figure 3.4. The MST Art method with edges as walls: (a) a stipple drawing; (b) the cycle formed by border edges; (c) a one-tree; and (d) the resulting maze.

approach, one must be able to stipple, and one must be able to solve TSPs! Only if we have the necessary ingredients (some stippling code and a TSP solver) can we use the TSP Art method to make mazes quickly and easily. With the MST Art approach, we are in better shape; anyone who has had just one semester of computer science should have no trouble producing a working version of Kruskal's algorithm. But to use the MST approach, we still need to be able to stipple.

Or do we?

Actually, we don't. If we allow ourselves to vary dot sizes and edge thicknesses, we can distribute our dots differently. One truly minimalist option is to use Vogel's model of phyllotaxis [14], the process by which plant leaves or

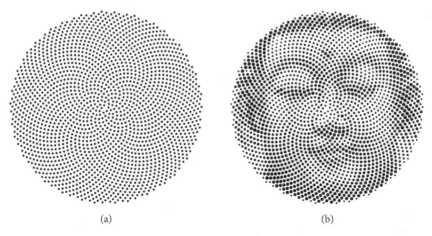

Figure 3.5. Two thousand dots (a) positioned using Vogel's model of phyllotaxis and (b) drawn with varying dot sizes.

seeds are arranged on their stem, to position the dots. For $k = 1, \ldots, n$ we form $\mathbf{z}_k = (x_k, y_k)$ with

$$x_k = C\sqrt{k}\cos(k\phi) \qquad \text{and} \qquad y_k = C\sqrt{k}\sin(k\phi),$$

where ϕ is the golden angle and C is a scaling constant. Figure 3.5a shows the result of positioning 2,000 dots in this way. Note that the dots form Fibonacci spirals: fifty-five moving inward in clockwise fashion, and eighty-nine moving inward in counterclockwise fashion (fifty-five and eighty-nine are consecutive Fibonacci numbers). Figure 3.5b shows the result of varying the sizes of these dots in accordance with the brightness levels of our target image.

Figure 3.6 displays phyllotactic TSP Art and MST Art mazes constructed from Figure 3.5b. These mazes look quite different from their nonphyllotactic counterparts. The phyllotactic mazes have what Xu and Kaplan [16] describe as directional texture; by focusing one's eyes on a small portion of these mazes, one finds that the pathways tend to run in the same direction. As a result, they may appear to look more maze-like. The nonphyllotactic mazes have what Xu and Kaplan describe as random texture. They look more dendritic; if one zooms into a small portion, one may be reminded of brain coral.

4 Seeded Stippling

To design mazes that have both the random textures of the original TSP Art and MST Art mazes and also the directional textures of the phyllotactic mazes, we developed a hybrid approach that employs both phyllotaxis and stippling. To explain it, we need to describe our approach to stippling.

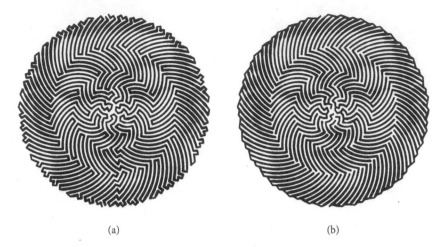

(a)   (b)

Figure 3.6. (a) A phyllotactic TSP Art maze. (b) A phyllotactic MST Art maze.

For most purposes, the best (and fastest) stippling algorithm is Secord's implementation [13] of Lloyd's method for constructing a weighted centroidal Voronoi tessellation [8]. The Secord/Lloyd method starts with an initial distribution of dots, and then, using the target image to assign a nonuniform density to two-dimensional space, repeatedly performs two operations: computing the Voronoi diagram of the dots, and moving each dot to the centroid (center of mass) of its Voronoi region. This easy-to-describe (but quite difficult-to-code) algorithm produces high-quality stipple drawings, drawings that have proved to be of tremendous value in the construction of high-quality pieces of TSP Art [10].

If speed is not an important consideration, there is a second option: MacQueen's method [12], a much easier-to-code (but much slower-to-converge) algorithm for constructing a weighted centroidal Voronoi tesselation. Bosch [2] used a modification of MacQueen's method to design simple-closed-curve sculptures of knots and links. Here is the algorithm:

0. Select an initial set of k dots $\{z_1, \ldots, z_k\}$ and set $n_i = 0$ for $i = 1, \ldots, k$.
1. Select a dot w according to a probability density function, derived from the brightness levels of the target image, that is more likely to pick dots that belong to darker regions of the target image than dots that belong to lighter regions.
2. Find the z_i that is closest to w.
3. Replace n_i with $n_i = n_i + 1$ and then replace z_i with

$$z_i = \frac{n_i z_i + w}{n_i + 1}.$$

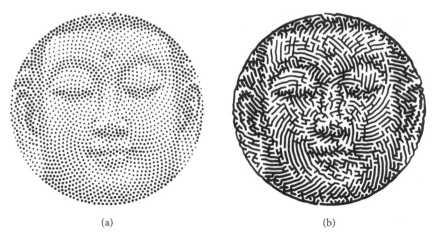

(a)   (b)

Figure 3.7. (a) A result of seeded stippling. (b) The corresponding MST Art maze.

4. If the new set of dots $\{z_1, \ldots, z_k\}$ satisfies some termination criterion, stop; otherwise, go to step 1.

Note that during each iteration, a dot w is used to *warp* (or move) one existing dot z_i. Exactly how much w warps the existing dot z_i depends on n_i, the number of times the existing dot has already been warped. The first time, the existing dot is moved halfway to the warping dot, the second time a third of the way, the third time a quarter of the way, and so on.

Our hybrid approach, which we call *seeded stippling*, simply replaces the initialization step of MacQueen's algorithm with the following:

0. Use Vogel's phyllotaxis model to generate a set of k dots $\{z_1, \ldots, z_k\}$ and set $n_i = 99$ for $i = 1, \ldots, k$.

Note that the initial configuration of dots comes from Vogel's model of phyllotaxis. Also, setting each n_i equal to 99 is done to *slow* the convergence of the algorithm. In the first iteration, the existing dot is moved a very small amount, just one one-hundredth of the way to the warping point. If this dot is moved on subsequent iterations, it will be moved even less. We want to move away from the phyllotaxis configuration, but not too far away from it.

Figure 3.7a displays the output of one run (of ten million small-step iterations) of our modified version of MacQueen's algorithm, and Figure 3.7d shows the corresponding MST Art maze. Note that this maze has, as desired, a mixture of random and directional textures.

Figure 3.8. Vortex tiles: (top) without connectors, (center) with the two possible horizontal connectors, and (bottom) with the two possible vertical connectors.

5 Vortex Tiles

Other minimalist approaches are possible. In this section, inspired by Xu and Kaplan [15], we describe how to form a maze by constructing an image mosaic out of the vortex tiles displayed in Figure 3.8. These vortex tiles are square tiles decorated with four polygonal paths and, if desired, up to four connectors. The polygonal paths start near the center of the square and spiral out to the midpoints of its sides. The greater the amount of spiraling, the darker the vortex tile will appear to the human eye. The connectors are horizontal or vertical line segments that connect points from which the spirals originate. The top row of Figure 3.8 shows that if no connectors are used, then a maze walker will be able to travel from any point in the tile to any other point in the tile without crossing a wall (a segment of a spiral). The middle row displays the tiles with the two possible horizontal connectors, and the bottom row displays them with the two possible vertical connectors. The middle and bottom rows show that if the two horizontal connectors or the two vertical connectors are used, then a maze walker will be able to go from either the top left corner to the bottom right corner or from the bottom left corner to the top right corner, but won't be able to do both.

Figure 3.9 illustrates the construction of a vortex maze. The first step is to impose a square grid on the target image (in this case, a small maze) and

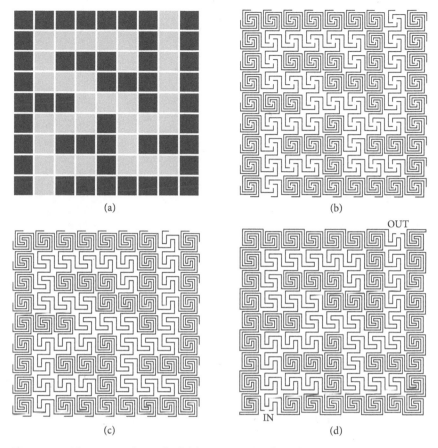

Figure 3.9. The vortex tile method (a) imposes a grid on the target image, (b) selects the amount of spiraling for each block, (c) selects the connectors, and (d) deals with the borders and picks an entrance and an exit.

measure the brightness of each block of pixels (by averaging the grayscale values of the pixels contained with the block). As shown in Figure 3.9a, we used a 9 × 9 grid. For this duotone target image, we ended up with dark blocks and light blocks.

The second step is to select the amount of spiraling needed for each block. Bright blocks need less spiraling, and dark blocks need more. As shown in Figure 3.9b, we used vortex tiles from the first and third columns of Figure 3.8. For the dark blocks, we used tiles from column three. For the light blocks, we used tiles from column one.

The third step is to choose which connectors to include at the centers of the tiles. For the mazes shown in Figures 3.9c and 3.10, we selected the

Figure 3.10. The Buddha (vortex-tile maze).

connectors in such a way as to make the walls of the mazes form simple closed curves.

The final step requires dealing with the dangling spiral segments on the borders and also picking an entrance and an exit.

6 Conclusion

Human beings have designed, studied, and played with mazes for more than 5,000 years [11]. In this chapter, we have demonstrated that simple mathematical methods can be used to design mazes that resemble user-supplied target images. We have provided enough details so that interested readers will be able to use these techniques to make their own mazes.

References

[1] D. L. Applegate, R. E. Bixby, V. Chvátal, and W. J. Cook. *The Traveling Salesman Problem: A Computational Study*. Princeton University Press, Princeton, NJ, 2006.

[2] R. Bosch. Simple-closed-curve sculptures of knots and links. *J. Math. Arts* **4** no. 2 (2004) 57–71.

[3] R. Bosch and A. Herman. Continuous line drawing via the Traveling Salesman Problem. *Op. Res. Lett.* **32** no. 4 (2004) 302–303.

[4] T. Chartier. *Math Bytes: Google Bombs, Chocolate-Covered Pi, and Other Cool Bits in Computing*. Princeton University Press, Princeton, NJ, 2014.

[5] W. L. Cook. *In Pursuit of the Traveling Salesman: Mathematics at the Limits of Computation*. Princeton University Press, Princeton, NJ, 2012.

[6] W. L. Cook, W. H. Cunningham, W. R. Pulleyblank, and A. Schrijver. *Combinatorial Optimization*. Wiley, New York, 1998.

[7] Concorde Home http://www.math.uwaterloo.ca/tsp/concorde.html (accessed September 10, 2014).

[8] Q. Du, V. Faber, and M. Gunzburger. Centroidal Voronoi tessellations: Applications and algorithms. *SIAM Rev.* **41** no. 4 (1999) 637–676.

[9] K. Inoue and K. Urahama. Halftoning with minimum spanning trees and its application to maze-like images. *Comp. Graphics* **33** (2009) 638–647.

[10] C. S. Kaplan and R. Bosch. TSP art, in R. Sarhangi and R. V. Moody, editors, *Proceedings of Bridges 2005: Mathematical Connections in Art, Music, and Science*, p. 301. Central Plains Book Manufacturing, Winfield, KS, 2005.

[11] H. Kern. *Through the Labyrinth: Designs and Meanings over 5000 Years*. Prestel, New York, 2000.

[12] J. MacQueen. Some methods for classification and analysis of multivariate observations, in L. M. Le Cam and J. Neyman, editors, *Proceedings of Fifth Berkeley Symposium on Mathematical Statistics and Probability, I*, p. 281. University of California Press, Berkeley, 1967.

[13] A. Secord. Weighted voronoi stippling, in *NPAR '02: Proceedings of the 2nd International Symposium on Non-photorealistic Animation and Rendering*, p. 37. ACM Press, New York, 2002.

[14] H. Vogel. A better way to construct the sunflower head. *Math. Biosciences* **44** no. 3–4 (1979) 179–189.

[15] J. Xu and C. S. Kaplan. Vortex maze construction. *J. Math. Arts* **1** no. 1 (2007) 7–20.

[16] J. Xu and C. S. Kaplan. Image-guided maze construction. *ACM Trans. Graphics* **26** no. 3 (2007) 29.

4

◇◇

SOME ABCs OF GRAPHS AND GAMES

Jennifer Beineke and Lowell Beineke

Dedicated to the memory of Martin Gardner on the occasion of the 100th anniversary of his birth

Martin Gardner had an exceptional gift for taking mathematical ideas that intrigued him and presenting them in a way that inspired interest in others. Since the National Museum of Mathematics hosted the conference that inspired this chapter, we are taking the view that the puzzles, games, and results we discuss are like objects in a museum. At its best, a visit to a museum leads individuals not only to appreciate what they see at the time, but to think about things afterward, perhaps even to create something of their own. Thus, it is in this fashion that we liken our collection to a return trip to a museum, where we look at some of our favorite graphical objects once again, perhaps seeing something we hadn't seen before, and also look at some items that are new or had gone unnoticed in earlier visits. We hope that readers will share in this experience and will be stimulated to investigate some of the ideas further.

In this chapter, we explore some mathematical exhibits in which graphs have a role, some explicit, some not obvious. We start with the Amazing Asteroid, following this with a theorem (Bernstein's Bijection), a couple of games (Chromatic Combat and Devious Dice), an episode of Eluding Execution, a coin-tossing game (Flipping Fun), and an African adventure game (Get the Giraffe). Versions of these explorations have all been successfully used with students at various levels—whether in the classroom, for Math Club, or for independent investigation—but they can be appreciated by wider audiences too. We conclude with a brief discussion of how one of the topics in particular is contributing to successful research experiences for undergraduates.

1 Amazing Asteroid

The first object in our selection of seven mathematical museum objects is the Amazing Asteroid, a game created by Jeremiah Farrell [10]. The game has two

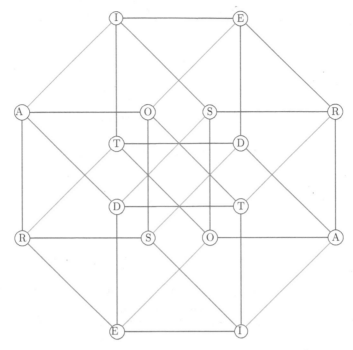

Figure 4.1. The magic tesseract in Amazing Asteroid.

players, Mind-Reader and Contestant. Contestant is asked to choose one of the eight letters in the word "ASTEROID" and also to decide whether to be a truth-teller or a liar. In front of Mind-Reader is not a crystal ball, but a magic tesseract (best if not visible to Contestant). It is labeled as shown in Figure 4.1, with each of the letters of "ASTEROID" appearing twice. In addition, the edges of the tesseract are colored with four colors.

Mind-Reader will ask Contestant four questions, each of which must be answered "yes" or "no." Moreover, Contestant must either answer all of them truthfully or all of them falsely.

The questions are these:

> Is your letter in the name Rosa?
> Is your letter in the name Dori?
> Is your letter in the name Rita?
> Is your letter in the name Otis?

As Mind-Reader, we start at the letter E (say, the one at the lower left) and move according to the "yes" answers. The key is that each of the four names is associated with one of the colors: Rosa with red, Dori with blue, Rita with green, and Otis with orange. For example, suppose that our player answered

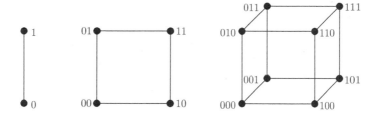

Figure 4.2. Zero-, one-, two-, and three-dimensional cubes.

the questions, "yes, yes, no, yes." From E, we go first to the right along the red edge to I, then we go vertically along the blue edge to T, then we don't take the green edge but next go along the orange edge to O.

We then claim that the letter that Contestant chose was O, and indeed O is in all the names except the third. Furthermore, from this, we can announce that Contestant was telling the truth. Note that if instead Contestant was a liar, the responses would have been "no, no, yes, no," and we would have gone up the green edge from E to the other O.

We now take a brief look at the mathematics behind the game. The tesseract is also known as the (4-*dimensional*) *hypercube*, one in a sequence of objects. The d-dimensional cube can be defined as follows: it has 2^d vertices labeled with the 2^d d-tuples of 0s and 1s, where two vertices are adjacent if their labels differ in exactly one position. The smaller cubes are shown in Figure 4.2, with the tesseract in Figure 4.3. Each of the thirty-two edges arises in one of the four dimensions. In our figure, this means that the first dimension is red (horizontal), the second blue (vertical), the third green (uphill), and the fourth orange (downhill).

Going from one vertex to another can therefore be described by a sequence of colors (directions). Thus, going from E to O, as in our example, can be described as "red, blue, and orange"; that is, by the first, second, and fourth dimensions. Of course the order of the colors doesn't matter; changing the order just means one takes a different path to arrive at the same point. Consequently, given any vertex and any subset of the four dimensions (or colors), one always has a well-defined ending position, and each of the sixteen vertices can be reached using one of the subsets. We also note that taking the complement of a set of colors results in arriving at the opposite vertex, the one that has its 0s and 1s interchanged but is labeled with the same letter.

To use this mathematics in an entertaining way, Farrell needed to—and was able to—find an eight-letter word (ASTEROID) in which there were four, four-letter words that contain the eight letters unambiguously (ROSA, DORI, RITA, OTIS). For example, the letter E is not in any of the names, and no letter is in all of them. The letter T is in only RITA and OTIS, and no other letter is in precisely

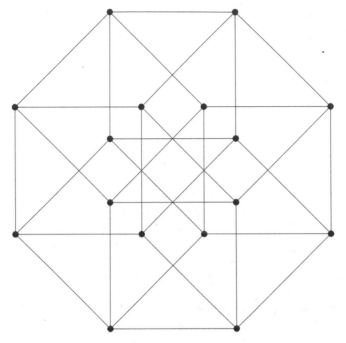

Figure 4.3. The 4-dimensional cube.

these two names or in just the other two, ROSA and DORA. This in itself was a remarkable achievement; indeed, quite amazing.

We hope that visitors to our small museum will enjoy amazing others with the wonderful magic tesseract.

2 Bernstein's Bijection

A favorite B exhibit for us involves the use of some graph theory to prove a classic result of set theory—Bernstein's theorem, also known as the Cantor-Bernstein theorem or the Cantor-Schroeder-Bernstein theorem.

Bernstein's Theorem. *If A and B are sets for which there is an injection from A into B and an injection from B into A, then there is a bijection between A and B.*

We note that consequences of this result include such facts as the existence of a one-to-one correspondence between the points of any open interval and the points of any closed interval.

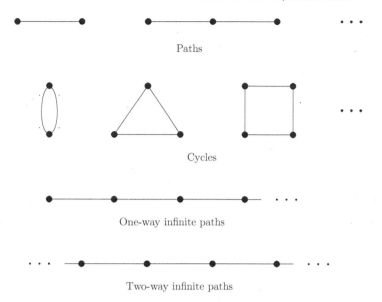

Paths

Cycles

One-way infinite paths

Two-way infinite paths

Figure 4.4. Graphs with maximum degree 2.

Although traditional proofs of Bernstein's theorem (see, e.g., [12], [19], and [13]) are not lengthy, they are quite technical, usually involving functions or sets defined recursively. At the end, students may not really understand why the result is true (although few would doubt it, even without a proof). The proof that we give is due to Julius Kőnig, whose son Dénes is perhaps more famous, being the author of the first book on graph theory, *Theorie der Endlichen und Unendlichen Graphen* ([14]). In fact, it is in that book (pp. 85–87) that Julius's proof appeared.

Before giving a proof of that theorem, we state the graph theory result that is the basis of its proof. This lemma needs no proof.

Lemma. *Every connected graph, finite or infinite, with maximum degree 2 is either a cycle or a path.* (See Figure 4.4.)

Proof of Bernstein's theorem. By hypothesis, A and B are two sets for which there exist injections $f : A \to B$ and $g : B \to A$. Without loss of generality, we assume that A and B are disjoint. We then form a bipartite graph G whose partite sets are A and B. We join a vertex $a \in A$ to a vertex $b \in B$ by a red edge if $b = f(a)$ and by a blue edge if $a = g(b)$ (see Figure 4.5). Note that every vertex in A is on a red edge and every vertex in B is on a blue one, with some vertices being on edges of both colors.

By the lemma, each component of G is either a cycle or a path. However, each of the cycles must be of even length. The only other possibility, for a

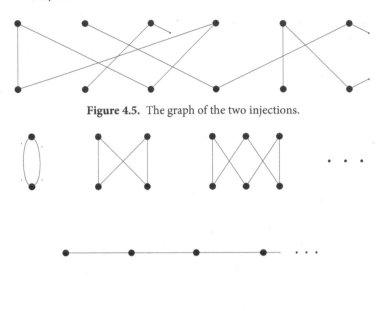

Figure 4.5. The graph of the two injections.

Figure 4.6. A bijection between A and B.

component with all vertices of degree 2 is a two-way infinite path. Now suppose some vertex, say $a_1 \in A$, has degree 1; that is, it is the end vertex of a path. It is therefore joined to a vertex b_1 by a red edge. But that vertex must be joined by a blue edge to another vertex $a_2 \in A$, which in turn is adjacent to some other vertex $b_2 \in B$. Note that since a_1 has degree 1, no vertex can be visited a second time. Hence this procedure must continue indefinitely, and from this it follows that every component with a vertex of degree 1 must be a one-way infinite path. Consequently, every component of G is either an even cycle or a one-way or a two-way infinite path. Clearly, any such subgraph has a perfect matching of its vertices (see Figure 4.6), and hence, so does G itself. □

3 Chromatic Combat

Our next museum piece has a different character; it's a game that we call Chromatic Combat, but it is also known as the Maker-Breaker game. Play involves coloring the vertices of a graph. It was first proposed by Steve Brams and appeared in one of Martin Gardner's Mathematical Games columns in *Scientific American* [11] in 1981. Investigation into the game for graphs in general began a decade later, first by H. Bodlaender [3] in 1991. Our description is based on the "museum piece" of Zsolt Tuza and Xuding Zhu [18].

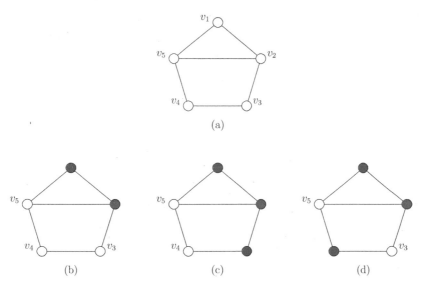

Figure 4.7. One game of Chromatic Combat.

Chromatic Combat is a two-player game; call the players Maker and Breaker. They are given a graph G and a set S of k colors, say $\{1, 2, \ldots, k\}$. Play alternates, with Maker going first, and a move consists in coloring a previously uncolored vertex with a color that none of its neighbors has (graph theorists refer to this as a proper coloring of a graph). Maker's goal is to achieve a proper coloring of the entire graph, while Breaker's goal is to make that impossible with the given set of colors.

For example, consider the graph G obtained from a 5-cycle by adding one chord, shown in Figure 4.7a, with three available colors, red, blue, and green. Play proceeds as follows: Maker colors vertex 1 red, with Breaker following by coloring vertex 2 blue, as in Figure 4.7b. If Maker then colors vertex 3 red (as in Figure 4.7c), then Breaker can color vertex 4 green, thereby blocking Maker from winning. However, if Maker were instead to color vertex 4 red (as in Figure 4.7d), nothing that Breaker does can prevent Maker from winning.

We observe that for there to be a game at all, k needs to be at least as large as the *chromatic number* $\chi(G)$ of the graph, the minimum number of colors for which there is a proper coloring. Clearly, if graph G has no more than k vertices, then Maker can always win by always playing an unused color, no matter what Breaker does. These observations suggest the following definition: The *game chromatic number* $\chi_g(G)$ of a graph G is the minimum number of colors for which Maker has a winning strategy no matter how Breaker plays. This is not only at least as large as $\chi(G)$, but is at most one more than the maximum number of neighbors of any vertex in G.

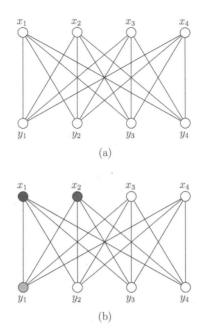

Figure 4.8. Chromatic Combat: play on $K_{4,4}$.

As another example, consider the complete bipartite graph $K_{4,4}$ with partite sets of vertices $X = \{x_1, x_2, x_3, x_4\}$ and $Y = \{y_1, y_2, y_3, y_4\}$ (shown in Figure 4.8a) and three colors, red, blue, and green. After Maker's first move, say, coloring x_1 red, Breaker's best move is equivalent to coloring x_2 blue, so we assume that is done. But then Maker can color y_1 green (see Figure 4.8b). Now the only possibilities are for Breaker either to color some uncolored vertex in X red or blue or to color some uncolored vertex in Y green. In either case, Maker can follow with whichever of these moves Breaker does not make. Continuing in this way, Maker will achieve a proper coloring of the graph. In contrast, it is easy to see that if there are only two colors available for this graph, then Breaker wins after the first round, so the game chromatic number is $\chi_g(K_{4,4}) = 3$.

Now consider the complete bipartite graph with a perfect matching removed, $CP_{4,4} = K_{4,4} - 4K_2$, sometimes called the "bipartite cocktail party graph" (there are four husband-wife couples at a party, and everyone talks to all those of the opposite sex other than their own spouse and talks to no one of the same sex). We assume the partite sets are as before but that for each i, x_i and y_i are not adjacent, as in Figure 4.9a. Also, the three colors, red, blue, and green, are again available.

Now, whatever vertex Maker colors (say, v_1 is colored red), Breaker will color the "spouse" of that vertex the same color (in this case, red). Thus, after two rounds of play, the situation is as in Figure 4.9b. It follows that Maker

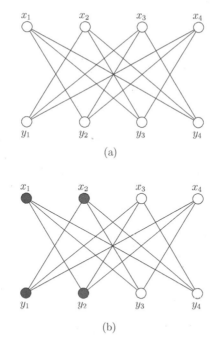

$$x_1 \quad x_2 \quad x_3 \quad x_4$$

$$y_1 \quad y_2 \quad y_3 \quad y_4$$

(a)

$$x_1 \quad x_2 \quad x_3 \quad x_4$$

$$y_1 \quad y_2 \quad y_3 \quad y_4$$

(b)

Figure 4.9. Chromatic Combat: play on $C P_{4,4}$.

will need four colors to achieve a proper coloring, so $\chi_g(C P_{4,4}) = 4$. These examples illustrate a couple of interesting features of the game chromatic number. First, its value for a subgraph of a graph can be greater than for the graph itself. Indeed, for the graphs $K_{n,n}$ and $C P_{n,n}$, the difference gets arbitrarily large. Another point that the bipartite cocktail party graph demonstrates is that it really matters who goes first: if Breaker goes first, then Maker can always win with just two colors.

Much more can be said and studied with regards to the Maker-Breaker game. An excellent resource, which discusses a variety of graph-coloring games and on which our description is based, is Tuza and Zhu [18], mentioned earlier.

4 Devious Dice

Under the Ds in the exhibit is a set of three colored dice with unconventional numbers, as shown in Figure 4.10:

Red: three 2s, two 11s, one 14
Blue: one 0, two 3s, three 12s
Green: three 1s, three 13s

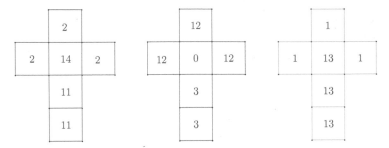

Figure 4.10. Devious Dice: Schwenk dice.

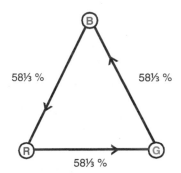

Figure 4.11. Devious Dice: probabilities in the one-die game.

These dice, the creation of Allen Schwenk [15], can be used to play a variety of games, the simplest of which is the one-die game: Player 1 chooses one of the three dice, after which Player 2 chooses one of the others; they then both roll their dice, and the one with the higher number wins. Given the two colors, is the game fair (under the assumption that the dice themselves are fair)? Suppose that Player 1 chooses red and Player 2 chooses green. The three 2s on the green die are losers no matter what Player 1 rolls, but the three 13s beat all but the 14 on the red die. Thus on average, green will beat red only fifteen times out of thirty six, or 5/12 of the time. Hence, the game is certainly not fair.

In fact, the same argument shows that blue will on average beat red 7/12 of the time. Therefore, if Player 1 chooses red, Player 2 should choose blue. But what should Player 2 do if Player 1 chooses the blue die? The calculations show that green beats blue the same fraction $(58\frac{1}{3}\%)$ of the time. Thus, paradoxically, blue is favored over red, red is favored over green, and green is favored over blue, all by the same amount. This paradox is summarized in Figure 4.11.

However, the story does not end there. Assume that there are two dice of each color and that each player throws both dice of some color. It seems reasonable to expect that, since a single blue die beats a single red die, if Player 1 chooses the pair of red dice, then Player 2 should choose the blue dice. One can compute that, in fact, a pair of red dice is better than a pair of blue ones

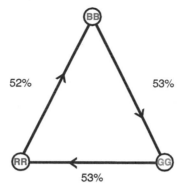

Figure 4.12. Devious Dice: probabilities in the two-dice game.

675 times out of the 1,296 possibilities, while a blue pair is better only 621 times. Thus, blue is preferable to red if only one die is rolled, but red is preferable to blue if two are rolled—a surprising reversal!

Even more astonishing is that the other two pairs of colors are also reversed. As already noted, a pair of reds is favored over two blues 675 to 621; additionally, a pair of blues is favored over two greens and a pair of greens over two reds, both by the slightly wider margin of 693 to 603. These facts are summarized in Figure 4.12.

Another possible two-dice game is with pairs of dice of any color combination. It then happens that a pair of Schwenk dice of one color against a mixed pair of the other colors (RR vs. BG, BB vs. GR, or GG vs. RB) is always a fair match. The probabilities (to two decimal places) of all other combinations are shown in Figure 4.13. We observe that each of the eight faces in this oriented graph is a 3-cycle. In fact, this is the most 3-cycles that any oriented graph on six vertices can have. There are also several 6-cycles, including this one:

$$RR \rightarrow GR \rightarrow GG \rightarrow BG \rightarrow BB \rightarrow RB \rightarrow RR.$$

I think you will agree with us that Schwenk dice are definitely devilishly devious.

5 Eluding Execution

Consider the following scenario. There are one hundred captives (all happen to be men), numbered from C_1 to C_{100}, in confinement, and either all will be executed or all will be freed. The deciding factor is as follows. There is a room with a hundred jars, also numbered from 1 to 100, and each contains the number of one of the captives (and each captive's number is in one of the jars). One by one, the captives are led into the jar room. Each captive can look inside

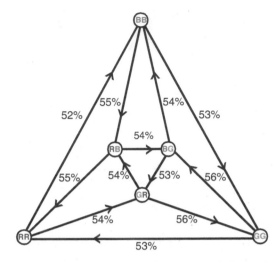

Figure 4.13. Devious Dice: probabilities for any two dice.

fifty of the jars to try to find his own number. He must leave the room exactly as he found it, and he is not allowed to communicate with any of the others afterward. The catch is that all of them must find their own number, or all will be executed.

Since each of them has probability $\frac{1}{2}$ of finding his number, and since their selections are independent of one another, the probability of their going free is $\frac{1}{2^{100}}$, or about

$$0.0000000000000000000000000000008.$$

However, there is, of course, more to the tale. The captives are allowed to confer beforehand. This they do, and they decide on the following strategy.

Captive C_1 goes to Jar 1. If Jar 1 contains the number 1, then C_1 is done. If not, then Jar 1 contains some other number, say i, and so C_1 goes to Jar i. He continues in this fashion until either he finds his own number or he has looked inside fifty jars, after which he leaves (and everyone will be executed).

Captive C_2 then goes into the room of jars and looks in Jar 2, using the same procedure that C_1 used. He is of course followed by all of the other captives in succession.

So what is the probability that the captives will all be executed in this scenario?

To start our analysis of the captives' approach to their dire situation, we observe that the numbers in and on the jars form a permutation π of the set $\{1, 2, \ldots, 100\}$, and we recall that a permutation can be considered as a labeled graph in which each component is a directed cycle. Another key observation

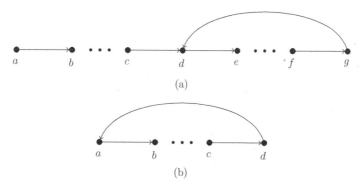

Figure 4.14. Eluding Execution: no jar is visited twice by one captive.

is that a captive never visits the same jar twice. For, suppose Jar k is visited twice by Captive j. Now if $j \neq k$ (see Figure 4.14a), then the number k would be in two jars, and this is impossible. In contrast, if $j = k$ (see Figure 4.14b), then the number j had to be in Jar j, and Captive j would have found his number at the start. We can therefore draw this conclusion: *if a captive does not find his number, then he must have started on a cycle of length greater than 50.*

That observation serves as a lemma to this theorem: *if the permutation π has no cycles of length greater than 50, then the captives elude execution.*

That raises a key question, which in a general form is this: what is the probability that a permutation of $2n$ objects has no cycles of length greater than n?

Let π now be a permutation of $2n$ objects, and assume that it has a cycle C of length k, with $k > n$. (Note that there can be at most one cycle whose length exceeds n.)

First we count the number of permutations of $2k$ objects:

    (a) The number of ways to choose the k objects in C is $\binom{2n}{k}$.

    (b) The number of ways to order the k objects in C cyclically is $(k-1)!$.

    (c) The number of ways to order the other $2n - k$ objects is $(2n - k)!$.

Next we multiply these three numbers and find that for a given value of $k > n$, the number of permutations of $2n$ objects that contain a k-cycle is

$$\frac{(2n)!}{k!(2n - k)!} \cdot (k - 1)! \cdot (2n - k)! = \frac{(2n)!}{k}.$$

Consequently, the probability of π having a k-cycle is just $\frac{1}{k}$, which leads to the following result.

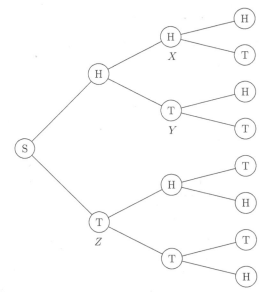

Figure 4.15. Flipping Fun: the tree of probabilities for flipping a coin.

Theorem. *The probability that a permutation of 2n objects has no cycles of length greater than n is*

$$1 - \left(\frac{1}{n+1} + \frac{1}{n+2} + \cdots + \frac{1}{2n} \right).$$

Since $1 + \frac{1}{2} + \cdots + \frac{1}{n}$ is approximately $\ln n$, it follows that $\frac{1}{n+1} + \frac{1}{n+2} + \cdots + \frac{1}{2n}$ is approximately $\ln(2n) - \ln n$, or $\ln 2$. Therefore the probability that the captives avoid execution is slightly better than 31%—quite a bit better than the figure $\frac{1}{2^{100}}$ cited earlier!

6 Flipping Fun

Another of our favorite objects in this collection is a classic coin-flipping game. Two players each choose a sequence of three in Heads and Tails, such as HHT and TTT. A coin is tossed until one of the chosen sequences appears in succession, and then the player who chose that sequence is the winner. For example, given the preceding choices, if the coin comes up TTHTTHTHTHHT, then the first player wins.

Naturally, it is possible for any sequence of three to beat any other triple—the first sequence might well come up as the first three tosses—but in our example, the second player could have made a choice that would have a better chance of winning, such as THH. This can be seen using the probability tree shown in Figure 4.15.

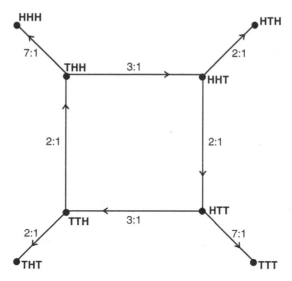

Figure 4.16. Flipping Fun: some odds in the game.

We derive the probability that Player 2 with THH will defeat Player 1 with HHT. Observe that every path in the tree will go through one of the three points labeled X, Y, and Z. If point X (that is, two Heads) is reached, then Player 1 will win (whenever the first Tail appears). In contrast, from either point Y or Z, Player 2 will win because, since a Tail has been tossed, THH is bound to appear before HHT (since two Heads have not yet occurred). Now the probability of reaching X is just $\frac{1}{4}$, as it is for reaching Y, while the probability of reaching Z is $\frac{1}{2}$. Hence the probability of Player 2 winning is $\frac{3}{4}$, and the odds favoring THH over HHT are 3 to 1.

As it happens, no matter what triple Player 1 picks, Player 2 can counter with one that has odds of at least 2 to 1 of winning, and they may be as high as 7 to 1. Optimal choices for Player 2 are shown in Figure 4.16, where the number $i : j$ on a directed arc is the odds of i to j in favor of the triple at the tail of the arc over the triple at the head.

We note that not only does the figure show that Player 2 has such a favorable strategy, but there is the paradox that THH has odds of 3 to 1 over HHT, which in turn has odds of 2 to 1 over HTT, and that has odds of 3 to 1 over TTH, yet the odds between THH and TTH are 2 to 1 *against* HHT!

7 Get the Giraffe

The last object in this tour of exhibits of graphs and games is Get the Giraffe. The game begins with a board with a 3×3 grid having an animal in each

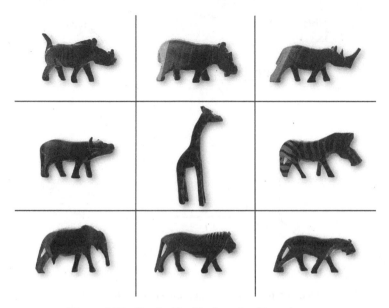

Figure 4.17. Get the Giraffe: the animal board game.

of the nine squares, as shown in Figure 4.17. Along with this is a "cage" to cover individual animals in the grid. (This could be a disk if the animals are two-dimensional or a cup in the three-dimensional case.) Also used is a set of eight cards, each with an instruction on one side to move a specified number of squares, and the name of one of the animals on the other side. Figure 4.18 shows an example of such a card.

The game has effectively just one player, whom we call Hunter. Play proceeds by Hunter, having put the cage over one of the animals, moving the cage the number of spaces mentioned on the card. Moving one space means going to one of the adjacent squares, but not diagonally. For example, in the board shown in Figure 4.17, if the lion is covered to start and the card says to make five moves, then Hunter could go from the lion to the giraffe to the zebra, back to the giraffe, up to the hippo, and finally ending on the rhino. Since the animal on the reverse side of the card is not the one caged, it is able to flee from Hunter, and hence leaves the board.

If at any time during the game, the animal named on the card is the one caged, then Hunter is the winner, having bagged a trophy. The game is over—and Hunter loses—if only one animal is left on the board.

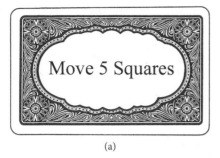 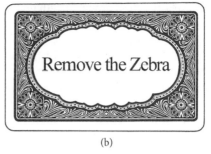

(a)   (b)

Figure 4.18. Get the Giraffe: a sample card. (a) Front of card. (b) Back of card.

One possible set of cards that achieves this outcome is the following, where the cage starts on the lion:

Card	Squares	Animal
1	5	Zebra
2	3	Warthog
3	4	Cheetah
4	3	Cape buffalo
5	6	Lion
6	3	Rhino
7	5	Hippo
8	1	Elephant

When this game is played, the giraffe is left alone on the board, and Hunter loses!

Why does Hunter lose? The answer lies in the pattern on the board. On a checkerboard, if the cage is moved an even number of steps, then it will be on a square of the same color as the one it started on, but if it is moved an odd number of steps, it will finish on a square of the opposite color. Consequently, once we have the color of the starting position settled, we control the game by specifying an animal on a color different from the one where the cage will stop. There is one detail missing here, however: we don't know where Hunter will decide to start at the beginning. To handle this, we simply have two first cards prepared, and then discreetly dispose of the one that won't work. (Some sleight of hand or feigned clumsiness may be useful here.)

Note that we could have used a board other than a checkerboard. The only requirement is that it correspond to a connected *bipartite* graph, one in which the vertices (the sites of the animals) are of two colors (like the black and white squares of a checkerboard) with no two vertices of the same color adjacent. For example, we could have a hyperspace version of the game using the tesseract (see the Amazing Asteroid section of this chapter).

We were introduced to this game by S. Brent Morris, who suggested one further twist: Before play starts, give a sealed envelope to a member of the audience. At the end of the game, that person opens the envelope to reveal the statement: The giraffe remains. (Why this animal? Well, after all, we are using *Giraffe Theory!*)

8 Hands On

Many modern museums, including the Museum of Mathematics, contain hands-on activities for visitors. In some cases, such participation will lead to questions that linger long after the visit. Whether it is an art museum, a natural history museum, a science museum, or the Museum of Mathematics, a visitor is likely not only to reflect on what was seen (and done), but also to think about the significance some of the items might have, paving the way for further questions and investigation.

Of the seven games and puzzles we have considered here, one stands out in this regard—Chromatic Combat. W. T. Tutte, one of the giants of graph theory, once wrote [17, p. 75] "The four color problem is the tip of the iceberg, the thin end of the wedge and the first cuckoo of spring." He was indeed prophetic, with hundreds of papers and several books having since been written on topics in chromatic graph theory.

An extremely fruitful area of research has been in variations of the chromatic number, one of which is the following. Relax the requirement that adjacent vertices always have different colors to allow each vertex to have one neighbor the same color, or more generally, up to some fixed number d of the same color. The *d-relaxed game* on a graph G is played by the same rules as Chromatic Combat, except for this additional freedom in coloring. That is, instead of each color generating a set of independent vertices, a color can give rise to a subgraph in which each vertex has degree at most d. This version of the game, introduced by Chou, Wang, and Zhu [4], has led to further research and interesting results, not just by mathematicians but by students too. For example, at Linville College, students have investigated relaxed colorings: see, for example, [1], [2], [5]–[9]. There are of course many relaxed coloring questions that are unanswered, but these students' work suggests other possible directions as well. What about other restrictions, such as list-colorings, in which each vertex can be colored only with a color from a specified list? (See Stiebitz and Voigt [16] for a survey on list-colorings.)

This is one example of how graphs and games create hands-on experiences that can be passed on to others. That thought brings us back to where we began our excursion—to Martin Gardner's great contributions. His ultimate achievement was the elegant, masterful way in which he handed on mathematical gems to others, setting an example for us all.

References

[1] M. Alexis, C. Dunn, J. F. Nordstrom, and D. Shurbert. Clique-relaxed coloring games on chordal graphs (in preparation).

[2] L. Barrett, C. Dunn, J. F. Nordstrom, J. Portin, S. Rufai, and A. Sistko. The relaxed game chromatic number of complete multipartite graphs (in preparation).

[3] H. L. Bodlaender. On the complexity of some coloring games. *Int. J. Found. Comp. Sci.* **2** (1991) 133–147.

[4] C.-Y. Chou, W.-F. Wang, and X. Zhu. Relaxed game chromatic number of graphs. *Discrete Math.* **262** (2003) 89–98.

[5] H. Do, C. Dunn, B. Moran, J. F. Nordstrom, and T. Singer. Modular edge-sum labeling (in preparation).

[6] C. Dunn, T. Hays, L. Naftz, J. F. Nordstrom, E. Samelson, and J. Vega. Total coloring games (in preparation).

[7] C. Dunn, V. Larsen, K. Lindke, T. Retter, and D. Toci. Game coloring with trees and forests. *Discrete Math. Theor. Comput. Sci.* (to appear).

[8] C. Dunn, D. Morawski, and J. Nordstrom. The relaxed edge-coloring game and *k*-degenerate graphs. *Order* (to appear).

[9] C. Dunn, C. Naymie, J. Nordstrom, E. Pitney, W. Sehorn, and C. Suer. Clique-relaxed graph coloring. *Involve* **4** (2011), 127–138.

[10] J. Farrell. Cubist magic, in D. Wolfe and T. Rodgers, editors, *Puzzler's Tribute, A Feast for the Mind*, A K Peters, Wellesley pp. 143–146, MA, 2002.

[11] M. Gardner. Mathematical games. *Sci. Am.* (April 1981) 23.

[12] C. Goffman. *Real Functions*, pp. 18–19. Rinehart, New York, 1953.

[13] M. S. Hellman. A short proof of an equivalent form of the Schroeder-Bernstein theorem. *Am. Math. Monthly* **68** (1961) 770.

[14] D. König. *Theorie der Endlichen und Unendlichen Graphen*. Chelsea Publishing, New York, 1950.

[15] A. J. Schwenk. Beware of geeks bearing grifts. *Math Horizons* **7** (2000) 10–13.

[16] M. Stiebitz and M. Voigt. List-colourings, in L. W. Beineke and R. J. Wilson, editors, *Topics in Chromatic Graph Theory*, pp. 114–136 Cambridge University Press, Cambridge, 2015.

[17] W. T. Tutte. Colouring problems. *Math. Intelligencer* **1** (1978/79) 72–75.

[18] Z. S. Tuza and X. Zhu. Colouring Games, in L. W. Beineke and R. J. Wilson, editors, *Topics in Chromatic Graph Theory*, pp. 304–326 Cambridge University Press, Cambridge, 2015.

[19] R. L. Wilder. *Introduction to the Foundations of Mathematics*, pp. 108–110. Wiley, New York, 1965.

PART II

Problems Inspired by Classic Puzzles

5

SOLVING THE TOWER OF HANOI WITH
RANDOM MOVES

Max A. Alekseyev and Toby Berger

The Tower of Hanoi puzzle consists of n disks of distinct sizes distributed across three pegs. We refer to a particular distribution of the disks across the pegs as a *state* and call it *valid* if on each peg the disks form a pile with the disk sizes decreasing from bottom up (so that the largest disk is on the bottom). Since each disk can reside on one of three pegs, while the order of the disks on each peg is uniquely determined by their sizes, the total number of valid states is 3^n.

At a single move it is permitted to transfer a disk from the top of one peg to the top of another peg, if this results in a valid state. In the puzzle's classic formulation, all disks are initially located on the first peg. It is required to transfer them all to the third peg with the smallest number of moves, which is known to be $2^n - 1$.

French mathematician Edouard Lucas invented the Tower of Hanoi puzzle in 1883 [6]. Apparently, he simultaneously created the following legend [7]: "Buddhist monks somewhere in Asia are moving 64 heavy gold rings from peg 1 to peg 3. When they finish, the world will come to an end!" That Hanoi is located in what was then French Indochina perhaps explains why a Frenchman saw fit to include Hanoi in the name of his puzzle. However, the legend never placed the monks and their tower explicitly in Hanoi or its immediate environs. Lucas's Tower of Hanoi puzzle became an international sensation (think Loyd's 15 puzzle and Rubik's cube); the legend was used to bolster sales. Still a popular and beloved toy, the Tower of Hanoi now also can be accessed over the Internet as a computer applet.

We shall study solutions of the Tower of Hanoi puzzle and some of its variants with random moves, where each move is chosen uniformly from the set of the valid moves in the current state. We prove exact formulas for the expected number of random moves to solve the puzzles. We also present an alternative proof for one of the formulas that couples a theorem about expected commute times of random walks on graphs with the delta-to-wye

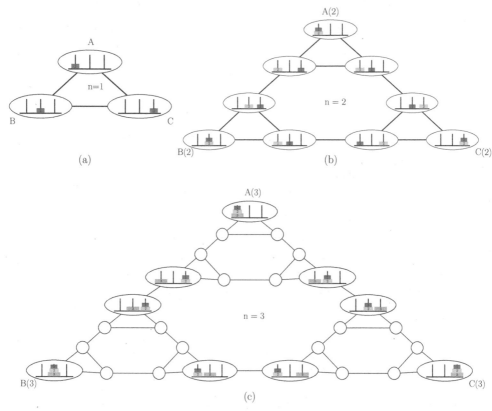

Figure 5.1. Sierpinski gaskets corresponding to (a) the state transition diagram for the one-disk Tower of Hanoi; (b) the state transition diagram for the two-disk Tower of Hanoi, which is composed of three replicas of the one-disk diagram with an added disk 2; and (c) the state transition diagram for the three-disk Tower of Hanoi, which is composed of three replicas of the two-disk diagram with an added disk 3 (only corner states in each replica are labeled).

transformation used in the analysis of three-phase AC systems for electrical power distribution.

1 Puzzle Variations and Preliminary Results

If the valid states of the Tower of Hanoi are represented as the nodes of a graph, and if every two states that are one move away from each other are connected with an edge, then the resulting graph is known as the Sierpinski gasket (Figure 5.1). In other words, the Sierpinski gasket represents the *state transition diagram* of the the Tower of Hanoi. More specifically, the Tower of Hanoi with $n = 1$ disk corresponds to a graph with three nodes A, B, and C and three edges $\overline{AB}$, $\overline{BC}$, and $\overline{AC}$ (Figure 5.1a). The nodes A, B, and C correspond, respectively, to the three possible states: D_1 is on the first peg, D_1 is on the second peg, and D_1 is on the third peg. The presence of edge $\overline{AC}$ represents the fact that it is possible to reach C from A or A from C in one move; the other two edges have analogous interpretations.

The graph for the state transitions of the Tower of Hanoi with $n = 2$ disks is obtained by arranging three replicas of the graph for $n = 1$ with an added disk D_2 at a fixed peg in each replica, and then connecting each of the resulting three pairs of nearest neighbor nodes by bridging links (Figure 5.1b). This leaves only three corner nodes, which we label $A(2)$, $B(2)$, and $C(2)$, such that $A(2)$ corresponds to both disks being on the first peg, $B(2)$ corresponds to both disks being on the second peg, and $C(2)$ corresponds to both disks being on the third peg.

Similarly, the graph for the state transitions of the Tower of Hanoi with $n = 3$ disks is obtained by arranging three replicas of that for $n = 2$, in the same way that was done to get the $n = 2$ graph from the $n = 1$ graph (Figure 5.1c). In general, for any positive integer k, the state transition diagram for $n = k + 1$ is obtained from that for $n = k$ by another iteration of this procedure employing three replicas and three bridges.

The classic Tower of Hanoi puzzle corresponds to finding the shortest path between two corner nodes in the corresponding Sierpinski gasket. We consider the following variants of the Tower of Hanoi puzzle with n disks solved with random moves (which correspond to random walks in the Sierpinski gasket):

> $r \to a$: The starting state is *random* (chosen uniformly from the set of all 3^n states). The final state is with all disks on the same (*any*) peg.
> $1 \to 3$: The starting state is with all disks on the *first* peg. The final state is with all disks on the *third* peg.
> $1 \to a$: The starting state is with all disks on the *first* peg. The final state is with all disks on the same (*any*) peg. At least one move is required.
> $1/2 \to a$: The starting state is with the largest disk on the *second* peg and the other disks on the *first* peg. The final state is with all disks on the same (*any*) peg.
> $r \to 1$: The starting state is *random* (chosen uniformly from the set of all 3^n states). The final state is with all disks on the first peg.

Let $E_X(n)$ denote the expected number of random moves required to solve Puzzle X with n disks. The puzzles described above are representative for classes of similar puzzles obtained by renaming pegs. In particular, we can easily get the following identities:

$$E_{1\to3}(n) = E_{1\to2}(n) = E_{2\to3}(n) = E_{3\to1}(n) = E_{2\to1}(n) = E_{3\to2}(n),$$

$$E_{1\to a}(n) = E_{2\to a}(n) = E_{3\to a}(n),$$

$$E_{1/2 \to a}(n) = E_{1/3 \to a}(n) = E_{2/3 \to a}(n) = E_{2/1 \to a}(n) = E_{3/1 \to a}(n) = E_{3/2 \to a}(n),$$

and

$$E_{r \to 1}(n) = E_{r \to 2}(n) = E_{r \to 3}(n).$$

Puzzle $r \to a$ was posed by David G. Poole, who submitted values of $E_{r \to a}(n)$ for n up to 5 as sequence A007798 to the Online Encyclopedia of Integer Sequences (OEIS) [9]. Later, Henry Bottomley conjectured the following formula for $E_{r \to a}(n)$:

$$E_{r \to a}(n) = \frac{5^n - 2 \cdot 3^n + 1}{4}. \tag{1}$$

Puzzle $1 \to 3$ was posed by the second author [1], who also submitted numerators of $E_{1 \to 3}(n)$ for n up to 4 as sequence A134939 to the OEIS but did not conjecture a general formula for $E_{1 \to 3}(n)$.

In Section 2, we prove formula (1). We also prove the following formula for $E_{1 \to 3}(n)$:

$$E_{1 \to 3}(n) = \frac{(3^n - 1)(5^n - 3^n)}{2 \cdot 3^{n-1}}, \tag{2}$$

which was originally announced by the first author in 2008. We also prove the following formulas for other puzzles:

$$E_{1 \to a}(n) = \frac{3^n - 1}{2}, \tag{3}$$

$$E_{1/2 \to a}(n) = \frac{3}{2}(5^{n-1} - 3^{n-1}), \tag{4}$$

and

$$E_{r \to 1}(n) = \frac{5^{n+1} - 2 \cdot 3^{n+1} + 5}{4} - \left(\frac{5}{3}\right)^n$$

$$= \frac{(3^n - 1)(5^{n+1} - 2 \cdot 3^{n+1}) + 5^n - 3^n}{4 \cdot 3^n}. \tag{5}$$

We summarize these formulas, and provide relevant references to the OEIS in Table 5.1.

2 Lemmas and Proofs

Without loss of generality, assume that the n-disk Tower of Hanoi has disks of sizes $1, 2, \ldots, n$. We denote the disk of size k by D_k, so that D_1 and D_n refer to the smallest and largest disks, respectively. Similarly, we let D_k^m ($k \geq m$) be the set of all disks of sizes from m to k inclusively.

TABLE 5.1.
Variations of the Tower of Hanoi puzzle with n disks

Puzzle X	Formula for $E_X(n)$	Initial values ($n = 1, 2, \ldots$)	Sequences in the OEIS
$r \to a$	$\frac{5^n - 2 \cdot 3^n + 1}{4}$	$0, 2, 18, 116, 660, \ldots$	A007798(n)
$1 \to 3$	$\frac{(3^n - 1)(5^n - 3^n)}{2} / 3^{n-1}$	$2, {}^{64}/_3, {}^{1274}/_9, {}^{21760}/_{27}, \ldots$	A134939(n) / A000244($n - 1$)
$1 \to a$	$\frac{3^n - 1}{2}$	$1, 4, 13, 40, 121, \ldots$	A003462(n)
$1/2 \to a$	$\frac{3 \cdot (5^{n-1} - 3^{n-1})}{2}$	$0, 3, 24, 147, 816, 4323, \ldots$	A226511($n - 1$)
$r \to 1$	$\frac{(3^n - 1)(5^{n+1} - 2 \cdot 3^{n+1}) + 5^n - 3^n}{4} / 3^n$	${}^4/_3, {}^{146}/_9, {}^{3034}/_{27}, {}^{52916}/_{81}, \ldots$	A246961(n) / A000244(n)

Note: For each Puzzle X, the table gives a formula for the expected numbers of random moves required to solve X with n disks, its numerical values for small n, and indices of the corresponding sequences in the OEIS [9].

In the solution of Puzzle $1 \to a$ with random moves, let $p_1(n)$ and $p_2(n)$ denote the probability that a final state has all disks on the first or second peg, respectively. From the symmetry, it is clear that the probability that a final state has all disks on the third peg is also $p_2(n)$, so $p_1(n) + 2p_2(n) = 1$.

Similarly, in the solution to Puzzle $1/2 \to a$ with random moves, let $q_1(n)$, $q_2(n)$, and $q_3(n)$ denote the probability that a final state has all disks on the first, second, or third peg, respectively.

The relationships of Puzzles $1 \to a$ and $1/2 \to a$ to Puzzles $r \to a$ and $1 \to 3$ are given by Lemmas 1 and 2 below.

Lemma 1.

$$E_{r \to a}(n) = E_{r \to a}(n - 1) + \frac{2}{3} E_{1/2 \to a}(n).$$

Proof. It is easy to see that D_n cannot move unless D_{n-1}^1 are on the same peg. In Puzzle $r \to a$, the expected number of random moves required to arrive at such a state is $E_{r \to a}(n - 1)$. Moreover, since the starting state is uniformly chosen, in the final state D_{n-1}^1 will be on any peg with equal probability $1/3$. In particular, with probability $1/3$ it is on the same peg where D_n resides and the puzzle is solved. Otherwise, with probability $2/3$ the disks D_{n-1}^1 and D_n are on distinct pegs, and thus we can view the remaining moves as solving an instance of Puzzle $1/2 \to a$. Therefore,

$$E_{r \to a}(n) = \frac{1}{3} E_{r \to a}(n - 1) + \frac{2}{3}(E_{r \to a}(n - 1) + E_{1/2 \to a}(n))$$

$$= E_{r \to a}(n - 1) + \frac{2}{3} E_{1/2 \to a}(n).$$

$\square$

Lemma 2.

$$E_{1\to 3}(n) = \frac{E_{1\to a}(n)}{p_2(n)}.$$

Proof. In the course of solving Puzzle $1 \to 3$ with random moves, all disks will first appear on the same peg after $E_{1\to a}(n)$ moves on average. This peg will be the first peg with probability $p_1(n)$, the second peg with the probability $p_2(n)$, or the third peg also with the probability $p_2(n)$. In the last case, Puzzle $1 \to 3$ is solved, while in the first two cases, we basically obtain a new instance of Puzzle $1 \to 3$. Therefore, $E_{1\to 3}(n) = E_{1\to a}(n) + (p_1(n) + p_2(n))E_{1\to 3}(n)$, implying that $E_{1\to 3}(n) = \frac{E_{1\to a}(n)}{p_2(n)}$. □

Lemmas 1 and 2 imply that explicit formulas for $E_{r\to a}(n)$ and $E_{1\to 3}(n)$ easily follow from those for $E_{1\to a}(n)$, $E_{1/2\to a}(n)$, and of $p_2(n)$.

Lemma 3. *The following equations hold:*

(i) $E_{1\to a}(n) = E_{1\to a}(n-1) + 2p_2(n-1)E_{1/2\to a}(n),$

(ii) $p_1(n) = p_1(n-1) + 2p_2(n-1)q_2(n),$

(iii) $p_2(n) \doteq p_2(n-1)q_1(n) + p_2(n-1)q_3(n),$

(iv) $E_{1/2\to a}(n) = \frac{1}{2} + E_{1\to a}(n-1) + (p_1(n-1) + p_2(n-1))E_{1/2\to a}(n),$

(v) $q_1(n) = p_1(n-1)q_1(n) + p_2(n-1)q_3(n),$

(vi) $q_2(n) = \frac{3}{4}(p_2(n-1) + (p_1(n-1) + p_2(n-1))q_2(n))$
$+ \frac{1}{4}(p_1(n-1)q_3(n) + p_2(n-1)q_1(n)),$ and

(vii) $q_3(n) = \frac{3}{4}(p_1(n-1)q_3(n) + p_2(n-1)q_1(n)) + \frac{1}{4}((p_1(n-1)$
$+ p_2(n-1))q_2(n) + p_2(n-1)).$

Proof. Consider Puzzle $1 \to a$. Note that D_n cannot move unless D_{n-1}^1 are on the same peg. Therefore, we can focus only on D_{n-1}^1 until they all come to the same peg, which will happen (on average) after $E_{1\to a}(n-1)$ moves. This will be the first peg (the location of D_n) with probability $p_1(n-1)$, in which case we have the final state with all disks on the first peg. Otherwise, with probability $1 - p_1(n-1) = 2p_2(n-1)$, we have D_n on the first peg and D_{n-1}^1 on a different peg (equally likely to be the second or the third), in which case the remaining moves can be considered as solving an instance of Puzzle $1/2 \to a$. This proves formula (i).

From the above, it is also easy to see that in the final state, all disks will be on the first peg with probability $p_1(n-1) + 2p_2(n-1)q_2(n)$ and on the second peg or third peg with the same probability $p_2(n-1)q_1(n) + p_2(n-1)q_3(n)$, which proves formulas (ii) and (iii).

Now consider Puzzle $1/2 \to a$. Moves in this puzzle can be split into two or three stages as follows. In Stage 1, only D_n is moving (between the second and third pegs), Stage 2 starts with a move of D_1 (from the top of the first peg) and ends when D^1_{n-1} are on the same peg. If this is not the final state, the remaining moves are viewed as Stage 3. Let us analyze these stages.

It is easy to see that the expected number of moves in Stage 1 is

$$\tfrac{2}{3} \cdot \left(\left(\tfrac{1}{3}\right)^0 \cdot 0 + \left(\tfrac{1}{3}\right)^1 \cdot 1 + \ldots \right) = \tfrac{1}{2},$$

and with probability

$$\tfrac{2}{3} \cdot \left(\left(\tfrac{1}{3}\right)^0 + \left(\tfrac{1}{3}\right)^2 + \ldots \right) = \tfrac{3}{4},$$

it will end up at the same peg where it started, namely the second peg. The probability for D_n to end up at the third peg is therefore $1 - \tfrac{3}{4} = \tfrac{1}{4}$. The expected number of moves in Stage 2 is simply $E_{1 \to a}(n-1)$. At the end, D^1_{n-1} are on the first peg with probability $p_1(n-1)$ and on the second or third pegs with equal probability $p_2(n-1)$. Therefore, with probability $p_2(n-1)$, we are at the final state (no matter where D_n is located after Stage 1), and with probability $1 - p_2(n-1) = p_1(n-1) + p_2(n-1)$, we embark on Stage 3, which can be viewed simply as a new instance of Puzzle $1/2 \to a$ with the expected number of moves $E_{1/2 \to a}(n)$. The above analysis proves formulas (iv)–(vii). $\square$

Formula (3) for $E_{1 \to a}(n)$ follows directly from (i) and (iv). Specifically, formula (iv) can be rewritten as $p_2(n-1)E_{1/2 \to a}(n) = \tfrac{1}{2} + E_{1 \to a}(n-1)$. Substituting this into formula (i) results in the recurrent formula:

$$E_{1 \to a}(n) = E_{1 \to a}(n-1) + 2p_2(n-1)E_{1/2 \to a}(n) = 3E_{1 \to a}(n-1) + 1.$$

Together with $E_{1 \to a}(1) = 1$ this result proves formula (3), which in turn further implies

$$E_{1/2 \to a}(n) = \frac{\tfrac{1}{2} + E_{1 \to a}(n-1)}{p_2(n-1)} = \frac{3^{n-1}}{2 p_2(n-1)}. \tag{6}$$

Let us focus on the recurrent equations (ii), (iii), (v), (vi), and (vii) and solve them with respect to $p_1(n)$, $p_2(n)$, $q_1(n)$, $q_2(n)$, and $q_3(n)$. Solving Puzzle $r \to a$ and Puzzle $1 \to 3$ for $n = 2$, we easily obtain the following initial conditions:

$$p_1(2) = 5/8, \quad p_2(2) = 3/16, \quad q_1(2) = 1/8, \quad q_2(2) = 5/8, \quad and \quad q_3(2) = 1/4.$$

Also, solving Puzzle $r \to a$ for $n = 1$, we get $p_1(1) = 0$ and $p_2(1) = 1/2$.

From equation (v) we have

$$p_2(n-1)q_3(n) = q_1(n) - p_1(n-1)q_1(n)$$
$$= (1 - p_1(n-1))q_1(n) = 2p_2(n-1)q_1(n).$$

Since $p_2(n-1) > 0$ for all $n \geq 2$, we also have

$$q_3(n) = 2q_1(n),$$
$$q_2(n) = 1 - q_1(n) - q_3(n) = 1 - 3q_1(n),$$

and

$$(8 - 3p_1(n-1))q_1(n) = 1$$

for all $n \geq 2$.

Using these relations, we simplify equation (ii) to

$$p_1(n) = p_1(n-1) + 2p_2(n-1)(1 - 3q_1(n)) = 1 - 6p_2(n-1)q_1(n)$$
$$= 1 - 3(1 - p_1(n-1))q_1(n) = 1 - 3q_1(n) + 3p_1(n-1)q_1(n)$$
$$= 1 - 3q_1(n) + 8q_1(n) - 1 = 5q_1(n).$$

Combining the above equations, we have $(8 - 15q_1(n-1))q_1(n) = 1$. That is,

$$q_1(n) = \frac{1}{8 - 15q_1(n-1)}. \tag{7}$$

Lemma 4. *For all positive integers n,*

$$q_1(n) = \frac{5^{n-1} - 3^{n-1}}{5^n - 3^n}. \tag{8}$$

Proof. We prove the formula for $q_1(n)$ by induction on n.

For $n = 1$, formula (8) trivially holds as $q_1(1) = 0$. Now for an integer $m \geq 1$, if formula (8) holds for $n = m$, then using formula (7), we get

$$q_1(m+1) = \frac{1}{8 - 15q_1(m)} = \frac{1}{8 - 15\frac{5^{m-1}-3^{m-1}}{5^m-3^m}}$$
$$= \frac{5^m - 3^m}{8(5^m - 3^m) - 15(5^{m-1} - 3^{m-1})} = \frac{5^m - 3^m}{5^{m+1} - 3^{m+1}}.$$

Therefore, formula (8) holds for $n = m + 1$, which completes the proof. $\qquad\square$

Formula (8) further implies:

$$q_2(n) = 1 - 3q_1(n) = \frac{2 \cdot 5^{n-1}}{5^n - 3^n},$$

$$q_3(n) = 2q_1(n) = \frac{2 \cdot (5^{n-1} - 3^{n-1})}{5^n - 3^n},$$

$$p_1(n) = 5q_1(n) = \frac{5^n - 5 \cdot 3^{n-1}}{5^n - 3^n},$$

and

$$p_2(n) = \frac{1 - p_1(n)}{2} = \frac{3^{n-1}}{5^n - 3^n}.$$

The last formula together with equation (6) proves formula (4).

Now we are ready to prove formulas (1) and (2). Lemma 1 together with $E_{r \to a}(0) = 0$ implies

$$E_{r \to a}(n) = \sum_{k=1}^{n} (E_{r \to a}(k) - E_{r \to a}(k - 1))$$

$$= \sum_{k=1}^{n} \left(5^{k-1} - 3^{k-1} \right) = \frac{5^n - 2 \cdot 3^n + 1}{4}.$$

Lemma 2 implies

$$E_{1 \to 3}(n) = \frac{E_{1 \to a}(n)}{p_2(n)} = \frac{(3^n - 1)(5^n - 3^n)}{2 \cdot 3^{n-1}}.$$

Finally, we derive formula (5). Solving Puzzle $r \to 1$ can be viewed as first solving Puzzle $r \to a$, and if our solution does not result in all disks on the first peg (which happens with probability $2/3$), continue solving it as Puzzle $1 \to 3$. Therefore, the expected number of moves in Puzzle $r \to 1$ is

$$E_{r \to 1}(n) = E_{r \to a}(n) + \frac{2}{3} E_{1 \to 3}(n) = \frac{5^n - 2 \cdot 3^n + 1}{4} + \frac{(3^n - 1)(5^n - 3^n)}{3^n}$$

$$= \frac{5^{n+1} - 2 \cdot 3^{n+1} + 5}{4} - \left(\frac{5}{3} \right)^n.$$

3 Analysis of Puzzle 1 → 3 via Networks of Electrical Resistors

We now present an altogether different method for solving Puzzle $1 \to 3$. This method relies on the puzzle's interpretation as a random walk between two corner nodes in the corresponding Sierpinski gasket and a result from electrical circuit theory.

The corner nodes of the Sierpinski gasket for the Tower of Hanoi with n disks correspond to the states with all disks on the first, second, or third peg, which we label $A(n)$, $B(n)$, and $C(n)$, respectively. In other words,

$$A(n) = (D_n^1, \emptyset, \emptyset), \quad B(n) = (\emptyset, D_n^1, \emptyset), \text{ and } C(n) = (\emptyset, \emptyset, D_n^1),$$

where $\emptyset$ is the empty set.

A *random walk* on an undirected graph consists of a sequence of steps from one end of an edge to the other end of that edge. If the random-walker currently is in a state S that has a total of M distinct states that can be reached from S in one step, then the random-walker's next step will go from S to each of these M states with probability $1/M$. Among the 3^n states of the Tower of Hanoi with n disks, $3^n - 3$ have $M = 3$; the other 3 (namely, the corner nodes $A(n)$, $B(n)$, and $C(n)$) have $M = 2$.

Building on a monograph by P. G. Doyle and J. L. Snell [3], A. K. Chandra et al. [2] proved the following theorem.

Theorem 1 (The Mean Commute Theorem). *The expected number of steps in a cyclic random walk on an undirected graph that starts from any vertex V, visits vertex W, and then returns to V equals $2m R_{VW}$, where m is the number of edges in the graph and R_{VW} is the electrical resistance between nodes V and W when a 1-ohm resistor is inserted in every edge of the graph.*

Figure 5.2a shows the graph for the one-disk Tower of Hanoi with a 1-ohm resistor inserted in each of its three edges. There are two parallel paths between states A and C, which are respectively the initial and final states for the Puzzle $1 \rightarrow 3$ with $n = 1$. The direct path along edge $\overline{AC}$ has a resistance of 1 ohm, and the indirect path along edge $\overline{AB}$ followed by edge $\overline{BC}$ has a resistance of 2 ohms. The overall resistance from A to C, therefore, is $1 \cdot 2/(1+2) = 2/3$ ohm. Since there are three edges in the graph, the mean commute time from A to C and back is $2 \cdot 3 \cdot 2/3 = 4$. By symmetry,[1] on average half of the time is spent going from A to C and the other half returning from C to A. Accordingly, the mean time it takes a Tower of Hanoi player making random moves with $n = 1$ to reach peg 3 starting from the first peg is $4/2 = 2$, which agrees with formula (2) for $E_{1 \rightarrow 3}(n)$ when $n = 1$.

We proceed to iterate this approach to derive formula (2) for general n. The key to performing the requisite iterations is the classical delta-to-wye transformation of electrical network theory [5]. A *delta* is a triangle with vertices A, B, and C that has resistances r_{AB} in edge $\overline{AB}$, r_{AC} in edge $\overline{AC}$, and

---

[1] Full symmetry is required to justify equal mean lengths of the outbound and return segments of a commute. For example, a 3-vertex graph with only two edges, $\overline{FG}$ and $\overline{GH}$, has full symmetry from F to H and back to F, but not from F to G and back to F. Simple calculations give $E_{FH} = E_{HF} = 4$, but $E_{FG} = 1$ whereas $E_{GF} = 3$.

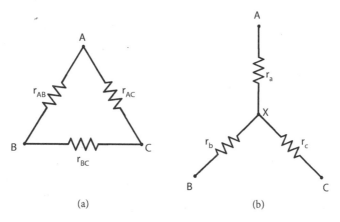

Figure 5.2. (a) Three-resistor delta; (b) Three-resistor wye.

r_{BC} in edge $\overline{BC}$ (Figure 5.2a). The corresponding *wye* (Figure 5.2b) has the same three nodes A, B, and C, plus a fourth node x and three edges $\overline{Ax}$, $\overline{Bx}$, and $\overline{Cx}$ that have resistances r_a, r_b, and r_c, respectively. It is straightforward to verify that, if

$$r_a = \frac{r_{AB}r_{AC}}{r_{AB}+r_{AC}+r_{BC}}, \quad r_b = \frac{r_{AB}r_{BC}}{r_{AB}+r_{AC}+r_{BC}}, \quad \text{and} \quad r_c = \frac{r_{AC}r_{BC}}{r_{AB}+r_{AC}+r_{BC}},$$

$$(9)$$

then the net resistance R_{AB} between nodes A and B will be the same in Figure 5.2b as it is in Figure 5.2a, and likewise for the net resistances R_{AC} between nodes A and C and R_{BC} between nodes B and C.

We need to consider only the special case

$$r_{AB} = r_{AC} = r_{BC} = R,$$

in which

$$r_a = r_b = r_c = R/3.$$

In particular, when $R = 1$, we have $r_a = r_c = 1/3$, so Figure 5.2b yields $R_{AC} = 2/3$, the same result we obtained before by considering the two parallel paths from A to C in Figure 5.2a.

Theorem 2 (Delta-to-Wye Induction). *The state diagram for the n-disk Tower of Hanoi with a unit resistance in each of its branches can be converted, for purposes of determining the resistance between any two of its three corner nodes $A(n)$, $B(n)$, and $C(n)$, into a simple wye in which $A(n)$, $B(n)$, and $C(n)$ each are connected to a center point by links that each contain a common degree of resistance denoted by $R(n)$.*

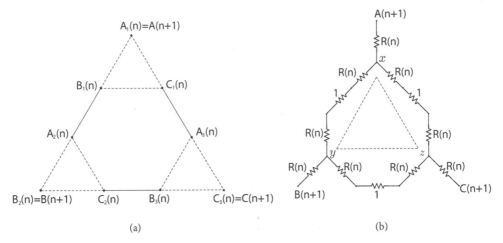

(a)   (b)

Figure 5.3. (a) State transition diagram; (b) the corresponding network of resistors for the Tower of Hanoi with $n + 1$ disks.

Proof. We prove Theorem 2 by induction on n. We have already shown that it is true for $n = 1$, the value of $R(1)$ being $1/3$ ohm. We now show that if Theorem 2 is true for some positive integer n, then it must also be true for $n + 1$.

As noted earlier, the state transition diagram of the Tower of Hanoi with $n + 1$ disks is produced by generating three replicas of that for the Tower of Hanoi with n disks that possess respective corner nodes

$$\{A_1(n), B_1(n), C_1(n)\}, \quad \{A_2(n), B_2(n), C_2(n)\}, \quad \text{and } \{A_3(n), B_3(n), C_3(n)\}$$

and then adding three bridging links: one between $B_1(n)$ and $A_2(n)$, another between $C_1(n)$ and $A_3(n)$, and the third between $C_2(n)$ and $B_3(n)$ (Figure 5.3a). The corner nodes in the resulting graph are

$$A_1(n) = A(n + 1), \quad B_2(n) = B(n + 1), \quad \text{and } C_3(n) = C(n + 1),$$

the only three nodes in the replicas to which none of the bridging links is incident. When applying Theorem 1, a unit resistance also must be inserted in each of the three bridging links, just as for every other link in the graph. By the induction hypothesis, the resistance between any two nodes in $\{A_1(n), B_1(n), C_1(n)\}$ can be computed using a wye comprised of links from each of them to a center point—call it x—each of these links having resistance $R(n)$. The same is true for any two nodes in $\{A_2(n), B_2(n), C_2(n)\}$ and any two in $\{A_3(n), B_3(n), C_3(n)\}$, with the respective center points called y and z. Doing these three delta-to-wye conversions results in Figure 5.3b. Note that triangle $\overline{xyz}$ in this figure is a delta, each edge of which has resistance

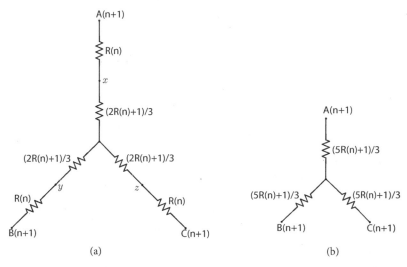

Figure 5.4. (a) A single wye of the state transition diagram of the Tower of Hanoi with $n + 1$ disks; (b) its reduction to a simpler wye with resistance $(5R(n)+1)/3$ in each link.

$R(n) + 1 + R(n) = 2R(n) + 1$. The delta-to-wye transformation applied to this delta network results in the wye network of Figure 5.4. Each of the links in this wye has the same resistance, namely,

$$R(n+1) = R(n) + \frac{2R(n) + 1}{3} = \frac{5R(n) + 1}{3}. \tag{10}$$

Theorem 2 is proved. □

From equation (10) and the boundary condition $R(1) = 1/3$, we obtain the key result:

$$R(n) = \frac{5^n - 3^n}{2 \cdot 3^n}. \tag{11}$$

The number m_n of edges in the state transition diagram for the n-disk Tower of Hanoi is

$$m_n = \frac{(3^n - 3) \cdot 3 + 3 \cdot 2}{2} = \frac{3}{2}(3^n - 1).$$

From the Mean Commute theorem and the symmetry of random walks from A to C and from C to A, it follows that the mean number of steps it takes a randomly moving n-disk Tower of Hanoi to transfer all its disks from the first peg to the third peg equals

$$m_n \cdot R_{AC}(n) = m_n \cdot 2R(n) = \frac{(3^n - 1)(5^n - 3^n)}{2 \cdot 3^{n-1}}, \tag{12}$$

which agrees with formula (2) for the mean number $E_{1 \to 3}(n)$ of moves in Puzzle $1 \to 3$.

4 Discussion

The minimum number of moves required to solve the Tower of Hanoi with $n = 64$ disks is "only" $2^{64} - 1 = 18,446,744,073,709,551,615$. Since it is often asserted that monks possess superhuman abilities, maybe they can move disks rapidly. Perhaps they can make a move a microsecond, maybe even a move a nanosecond, and planet Earth may expire any day now. This in part motivated the adoption of a randomly moving Tower of Hanoi [1].

Formula (12) shows that replacing the minimum-moves strategy with a random walk forestalls the end of the world by a factor of roughly $\left(\frac{5}{2}\right)^{64} > 2.9 \times 10^{25}$ on average. Although this is reassuring, it would provide further comfort to know that the coefficient of variation of the random number of steps in Puzzle 1 $\rightarrow$ 3 with $n = 64$ disks is small (i.e., that its standard deviation is many times smaller than its mean). Exact determination of said coefficient of variation is an open problem that we may address in future research.

An extensive bibliography of some 370 mathematical articles concerning the Tower of Hanoi puzzle and variations thereon has been complied by Paul Stockmeyer [8]. While the current paper was under review, our attention was drawn to work by Wu, Zhang, and Chen [10], which develops similar ideas of analyzing random walks in Sierpinski gaskets via resistor networks. Properties of the generalized Tower of Hanoi with more than three pegs and its state transition diagrams are studied to some extent in [4].

Acknowledgments

We are indebted to Neil J. A. Sloane, creator and caretaker of the Online Encyclopedia of Integer Sequences (OEIS) [9]. Neil's interest in the research reported here and the existence of his OEIS connected us to one another and to the broader Tower of Hanoi research community. We are also thankful to Sergey Aganezov and Jie Xing for their help with preparation of the figures.

The first author was supported by the National Science Foundation under grant No. IIS-1462107.

References

[1] T. Berger. Lucas, Sierpinski, Markov, Shannon and the end of the world, Invited Presentation in the David Slepian Memorial Session, Information Theory and Its Applications (ITA 2008), University of California, San Diego, January 2008.

[2] A. Chandra, P. Raghavan, W. L. Ruzzo, R. Smolensky, and P. Tiwari. The electrical resistance of a graph captures its commute and cover times, in *Proceedings of the*

21st Annual ACM Symposium on the Theory of Computing, Seattle, May 1989. ACM Press, New York, 1989.

[3] P. G. Doyle and J. L. Snell. *Random Walks and Electrical Networks*. Mathematical Association of America, Washington, DC, 1984.

[4] A. M. Hinz, S. Klavžar, U. Milutinović, and C. Petr. *The Towers of Hanoi—Myths and Maths*. Birkhäuser, Basel, 2013.

[5] A. E. Kennelly. Equivalence of triangles and stars in conducting networks. *Electrical World Engineer* **34** (1899) 413–414.

[6] E. Lucas. *Recreations Mathematiques, Volume III*, Gauthiers-Villiars, 1893. Reprinted by Albert Blanchard, Paris.

[7] R. L. Ripley. *The New Believe It or Not Book—2nd Series*. Simon and Schuster, New York, 1931.

[8] P. K. Stockmeyer. The Tower of Hanoi: A bibliography, Version 2.2, 2005. http://www.cs.wm.edu/~pkstoc/biblio2.pdf.

[9] The OEIS Foundation. *The Online Encyclopedia of Integer Sequences*, 2015. http://oeis.org.

[10] S. Wu, Z. Zhang, and G. Chen. Random walks on dual Sierpinski gaskets. *Euro. Phys. J. B* **82**, no. 1 (2011) 91–96.

6

GROUPS ASSOCIATED TO TETRAFLEXAGONS

Julie Beier and Carolyn Yackel

What do you do when you have paper that is too large for your notebook? If you are mathematician Arthur Stone, you cut strips off of your paper to make it fit and use the strips to create fascinating objects called *flexagons*. Developed in 1939, flexagons are typically flat folded paper objects with hidden faces. The folds act as hinges that must be simultaneously manipulated along multiple axes to shuffle the faces of the flexagon. This process, dubbed a "flexing action," hides the original object faces, bringing others to view. The most effective way to understand a flexagon is to build one and practice flexing it. Readers who have not had this experience are encouraged to either follow the directions in [2] or [4] to make the popular tri-hexaflexagon, or to watch the video by Vi Hart [3]. Since flexing acts nicely on a physical object, we are naturally led to wonder: Does the set of flexes form a group? In fact, we ask: If we include the dihedral symmetries corresponding to the flexagon shape with the flex actions, do we obtain a group?

The study of these questions began with the tri-hexaflexagon, a flexagon built with triangles in the shape of a hexagon with three sides. In 1972, Jean Pederson pointed out that the tri-hexaflexagon had a group of symmetries isomorphic to the integers modulo three [6]. However, in a 1997 *Mathematics Magazine* paper with Hilton and Walser, she expanded this initial group to the largest such group for the tri-hexaflexagon. This new group had thirty-six symmetries and proved to be the dihedral group on eighteen elements [4]. The hexa-hexaflexagon was the next flexagon whose group structure was explored. This flexagon again is built with triangles and is in the shape of a hexagon, but it has six sides, making it more complex. Berkove and Dumont published a 2004 article, also in *Mathematics Magazine,* that showed hexa-hexaflexagons do not admit a group structure on their set of actions, at least with the group associated to the flexagon in the way it had been done in the past [1]. With this method, Berkove and Dumont push further to indicate that most flexagons also fail to admit a group structure.

In response to the shocking negative result of Berkove and Dumont, we decided to look at the way groups are associated to flexagons by focusing on the mathematical encoding of the actions themselves. For instance, we study a simple tetraflexagon, which is square-shaped, that Berkove and Dumont construct by folding a lightning bolt–shaped net of squares [1]. They indicate its associated group to be the integers modulo four. We have found the full action group of this tetraflexagon by using techniques similar to Hilton, Walser, and Pederson, but by using a different method for mathematically encoding the actions than that of Berkove and Dumont. This group of actions has order much larger than four. This is as expected, since the symmetries of the square are added, but it is also surprisingly complicated!

In this chapter, we present this example, along with several other groups admitted by the actions of tetraflexagons. These tetraflexagons yield rich group structures that are unexpected from such simple objects. Throughout this chapter, we take a generators and relations approach to determining the appropriate group corresponding to each flexagon. For each example, we present the structure diagrams, discuss our choices in mathematizing, compare our choice to the previous literature, and report the associated group. Each example employs slightly different choices in the generator assignment, which serves to capture different aspects of the physical nature of the flexagon itself. While we have already uncovered some beautiful structure, there are still a number of difficulties realizing the groups for many tetraflexagons. We point out these challenges in hopes that the conversation will continue beyond this chapter.

1 Flexagon Basics

Flexagon is a term for any object in the family of objects folded from paper nets, which when folded and opened properly yield previously hidden sides. For details about the requirements to be a flexagon, see Pook's book [7]. The most commonly known example is the tri-hexaflexagon. These are triple-twist Möbius bands that have been flattened to a hexagonal shape. As mentioned earlier, they can be made simply from a strip of paper that has been folded into triangles. The prefix "tri" in the name "tri-hexaflexagon" signifies that the flexagon itself has three faces (one that is hidden until after the flex) and the "hexa" indicates the hexagonal shape. Directions for constructing a tri-hexaflexagon are available in the aforementioned references. In this chapter, we restrict our conversation to the less famous square-shaped tetraflexagon family.

In general, building a basic flexagon is easily accomplished when given folding directions and the shape of a *net*, or a two-dimensional connected arrangement of polygons that will be folded to create the flexagon. To make folding directions easier, we label each square as illustrated in Figure 6.1. Since orientation is rarely preserved with a flexagon, it is important to indicate the

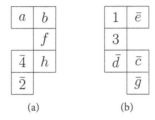

Figure 6.1. (a) Small snake net and (b) its reflection.

direction of the labels. The symbol "–" will indicate the baseline for the text if it is not as printed. For instance, "A" is as printed, while "$\overline{A}$" indicates an upside-down A, and "$|A$" indicates an A that has been rotated 90° clockwise. All labels in the nets in this paper begin either oriented with the standard baseline or upside-down, as seen in Figure 6.1. The labels may seem strange at this point, but the choice will be clear after the flexagon is assembled.

Authors employ a variety of methods to communicate how to fold and tape a flexagon from a given net. Most directions indicate the folding in steps to be completed. However, we opt for a different technique that clearly encodes a final folding of the flexagon itself instead of listing the steps. A finished tetraflexagon will be in the shape of a square with four quadrants. Each quadrant is called a *pat* and, in this chapter, pats on the same diagonal have the same thickness, as assumed in [1]. To each quadrant we associate a string that gives the configuration of the pat. A square in the net will be denoted by $\dfrac{p}{q}$, where p is the label on the square that is showing on top, and q is the label that is below. The symbol ">" indicates that the square on the left is on top of the square on the right. For instance, $\dfrac{a}{\bar{e}} > \dfrac{g}{2}$ indicates that the square with an a on top and an upside-down e on the bottom is on top of the square with g on the top and 2 on the bottom. Thus, when starting with the net oriented as in Figure 6.1a, to create this flexagon it is necessary to fold the square with $\bar{g}$ up (making it a g) and align it on top of the square labeled $\bar{e}$. Now the flexagon must be flipped over so that g is below $\bar{e}$, rather than above. This sequence will be given for each quadrant Q in the order $Q2 \diamond Q1 \diamond Q3 \diamond Q4$. In most cases this will be the only information given for a folding, leaving a small puzzle for the folder to solve.

Example 1. *The folding for the net in Figure 6.1 is*

$$\frac{a}{\bar{e}} > \frac{g}{2} \diamond \frac{b}{1} \diamond \frac{c}{4} \diamond \frac{d}{h} > \frac{f}{3}.$$

To understand this case clearly, observe that the easiest way to obtain this simple flexagon is by (i) mountain folding along the line between square d and 3, which

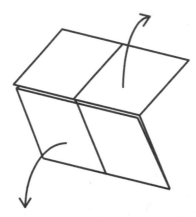

Figure 6.2. An x-down flex. To flex, hold the flexagon in quadrants Q1 and Q3 rotating the hand holding quadrant Q1 away from you and the hand holding quadrant Q3 toward you. Newly made flexagons may stick; pull firmly.

places square $\bar{h}$ on square f and then (ii) tucking the square labeled g under the square labeled a, placing $\bar{e}$ on top of g. Finally, the top of square a is taped the the top of the square below it. An expert flexagoner will notice that this is the only place to tape and will create an object that both stays together and flexes.

Notice that the one-line notation we are using gives more information than just the folding itself, which is why it is particularly useful in investigations.

Now that we have a flexagon, we must flex it to find the hidden sides. The flexagon that results from the directions given in Example 2.1 (oriented such that a, b, c, and d are on the top) can be flexed only one way. Viewing the pats as quadrants, there is an x-axis horizontally bisecting the flexagon and a y-axis bisecting vertically. Fold the flexagon down around the x-axis, as illustrated in Figure 6.2. Now open the flexagon to reveal a new side, labeled $efgh$ (reading Q2, Q1, Q3, Q4). We call this kind of flex an x-down flex. If you try to perform another x-down flex to your flexagon you will see that it can not be done because squares f and h are joined together. When two squares are joined in this way they form what is called a *bridge*. However, it is possible to complete an x-up flex from the current position of this flexagon. An x-up flex folds up around the horizontal centerline rather than down. Doing this flex will return your flexagon to its initial state, $abcd$. Note that some articles refer to these flexes as laptop flexes [1]. Often, tetraflexagons also admit flexes along the y-axis. We call these y-down and y-up flexes, and they work similar to the x-flexes. Observe that the flexagon from Example 1 fails to permit any y-flexes. These y-flexes are sometimes called book flexes [1]. You may notice that flexing changes the thickness of the pats in a predictable way. Therefore, a flex is not a true symmetry of the flexagon. Instead, we discuss the group of

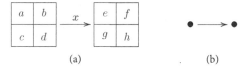

Figure 6.3. (a) Small snake structure diagram and (b) simplified diagram.

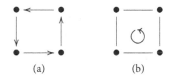

Figure 6.4. (a) Box structure diagram and (b) reduced diagram.

actions on the flexagon, where actions may consist of symmetries, flexes, or compositions of the two.

It is useful to keep track of how flexes move between the states of a flexagon. To do this, we create a *structure diagram*. The nodes of this diagram are the faces of the tetraflexagon that can be reached from the starting position using flexes. Arrows indicate down flexes that terminate at the resultant state. When possible, we use horizontal arrows for x-flexes and vertical arrows for y-flexes. See Figure 6.3a for the structure diagram of the flexagon made in Example 1. It is frequently helpful to suppress some of the detail in these diagrams. The reduced diagram is called the *simple structure diagram* and is formed by hiding the x and y labels, and replacing the flexagon states with nodes. This is demonstrated in Figure 6.3b.

There are times when a structure diagram may form a complete square, as illustrated in Figure 6.4a. In this case, we further simplify the diagram by removing all arrow heads and placing a curved arrow in the center of the diagram to indicate the direction that is traveled using down flexes.

The remainder of this chapter is devoted to examples and their groups. We begin with the most basic of tetraflexagons.

2 A Simple Case

The simplest tetraflexagon, which we call the "small snake," is depicted in Figure 6.2 and elaborated on in Example 1. It can also be found in Pook's book [7, Fig. 6.4]. The astute folder will realize that there exists another tetraflexagon built from the same net that behaves identically to this one. The two foldings are mirror images. Flexagons with this relationship are called *enantiomorphic*. The structure diagram for the flexagon in Example 1 is shown in Figure 6.3.

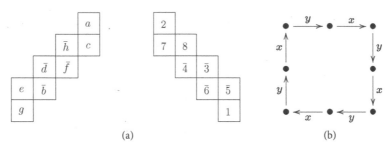

Figure 6.5. (a) Labeled stairstep net and (b) structure diagram.

We now seek to understand the group of symmetries of this flexagon. From any state of the flexagon, exactly one flex can be performed. Here let z denote a group element corresponding to any flex. Then we see that $z^2 = e$, since $z(abcd) = efgh$ and $z(efgh) = abcd$. Notice that $z^2 = e$ implies $z = z^{-1}$, which makes sense in this flexagon, since any flex can be undone by simply flexing again. The careful reader will notice that the square sides labeled 1, 2, 3, and 4 have yet to appear. To see these hidden squares, it is necessary to flip the flexagon over. Here, and for the remainder of the chapter, we choose to perform the flip across the y-axis and denote it as f. Notice that indeed, $f(abcd) = 1234$. This emphasizes the need to include

$$D_4 = \langle r, f \mid f^2 = r^4 = e, \, frf = r^3 \rangle$$

in the group of actions. Observation demonstrates that f and r, which is taken to be 90° counterclockwise rotation, commute with z. Thus, when mathematizing the actions in this way, the group of symmetries of the small snake is $\mathbb{Z}_2 \times D_4$.

3 Sub-Squares Matter

We now consider an interesting stairstep net from Pook's book [7, Fig. 6.1], which is given with labeling in Figure 6.5a. The folding for this net is given by

$$\frac{a}{2} \diamond \frac{b}{6} > \frac{5}{\bar{e}} > \frac{g}{1} \diamond \frac{c}{7} > \frac{8}{\bar{h}} > \frac{f}{4} \diamond \frac{d}{3}$$

(tape squares 1 and 2 along the upper y-axis). Its simplified structure diagram is shown in Figure 6.5b.

The fact that the diagram is a cycle brings immediate pleasure, because it seems that the flexagon is likely to have a group structure. In Pook's book, he claims that this flexagon has four faces that form a four cycle. To see this, imagine coloring each face one color. For instance, color the face that contains a, b, c, and d blue; the next face that contains 1, 2, 3, and 4 orange; and so on. In this way, four colors will be needed and thus four faces. For this flexagon, it is true that only one down flex can be performed from each state. From now on, we may use the notation z to indicate down flexes when there is only one possible down flex from each state. Then it is obvious that $z^4 = e$, and the group is $\mathbb{Z}_4$.

Closer inspection reveals more nuance. Start with the state $abcd$, and perform four flexes to obtain $\bar{d}\bar{c}\bar{b}\bar{a}$. While we see this state would still be colored blue, it is not the one with which we started. This indicates the importance of paying attention to the arrangement of the sub-squares. Observing sub-square arrangements shows there are actually eight faces, as suggested by the structure diagram. Thus, we have that $z^8 = e$ and may be falsely tempted to assume our group is $\mathbb{Z}_8 \times D_4$, or $\mathbb{Z}_8 \rtimes D_4$, depending on whether or not the elements of D_4 commute with $\mathbb{Z}_8$.

Detailed investigation reveals that some of the expected relations indeed hold, such as $fr = r^3 f$, as do some less obvious ones, including $rz = zr$, $z^4 = r^2$, $fz = z^7 f$, and $zf = z^4 f z^3$. The relation $z^4 = r^2$ is particularly significant, since it demonstrates that the generators of D_4 are intimately connected with the element z. To reveal this flexagon's group, carefully choose the generators rz^2, z^3, and f, rather than the standard generators. The order of these elements are 2, 8, and 2, respectively. Now, f commutes with rz^2 but $fz^3 = z^5 f = (z^3)^7 f$. Since f does not commute with the generator z^3, the direct product is not the appropriate way to join the groups $\langle z^3 \rangle$ and $\langle f \rangle$. Fortunately, the semidirect product, denoted $\rtimes$, permits the groups to join in a noncommuting manner. Recall that $G \rtimes_\phi H$ consists of the ordered pairs (g, h), with $g \in G$ and $h \in H$ and with multiplication defined so that $(g_1, h_1)(g_2, h_2) = (g_1 \phi(h_1)(g_2), h_1 h_2)$, where $\phi : H \to Aut(G)$ [5]. Hence, the group of actions here is unexpectedly $\mathbb{Z}_2 \times (\mathbb{Z}_8 \rtimes_\phi \mathbb{Z}_2)$, where the homomorphism $\phi : \mathbb{Z}_2 \to Aut(\mathbb{Z}_8)$ is given by defining $\phi(1)$ to be the automorphism of $\mathbb{Z}_8$ that sends 1 to 7.

4 Not All Flexes Are Equal

The next flexagon example has a straightforward bridge configuration, since the pairs of bridges are not intertwined. This example appears in Berkove and Dumont [1, Fig. 10] and Pook [7, Fig. 3.9]. We illustrate it together with its structure diagram in Figure 6.6. The net folding is given by

$$\frac{a}{7} > \frac{\bar{5}}{\bar{e}} \diamond \frac{b}{2} > \frac{1}{\bar{f}} \diamond \frac{c}{3} > \frac{4}{\bar{g}} \diamond \frac{d}{6} > \frac{\bar{8}}{\bar{h}}$$

(tape the right side of b to $\bar{f}$).

Again, the structure diagram is simply a cycle. Treating all down flexes as equal, as in the previous example, yields $z^4 = e$, verifying Berkove and Dumont's claim that the flexes give the group $\mathbb{Z}_4$. However, we notice that

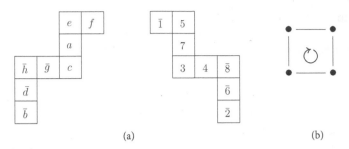

Figure 6.6. (a) Labeled lightning bolt net and (b) its structure diagram.

denoting all flexes as z alone obscures the physical difference between the horizontal and vertical flexes. In addition, we would like to enlarge the group to include the symmetries from D_4.

Assigning x and y as group generators for the corresponding x-down and y-down flexes causes closure to be a problem. As in the previous examples, only one down flex can be performed from any particular state. In addition, none of these flexes can be performed more than once in a row. Thus the relations $x^2 = e = y^2$ are natural to impose. The identifications $x = x^{-1}$ and $y = y^{-1}$ result. However, these are not unreasonable, since the inverse of x-down is x-up, and the inverse of y-down is y-up, as we observed in the small snake example. Hence the action group, without the D_4 symmetries, is generated by x and y, subject to the relations $x^2 = y^2 = e = (xy)^2$. The relation $(xy)^2 = e$ can be read from the structure diagram. A measured look at this relation shows that $xy = yx$, revealing the subgroup generated by x and y to be isomorphic to the Klein-4 group.

We note that in this flexagon, flexing does not mix the sub-squares. Thus, unlike in the stairstep example, the inclusion of D_4 does not cause any of the relationships with x and y to collapse. Observe that f commutes with both x and y, implying that conjugation by f is trivial. We highlight for the reader that on one side of the equation $fx = xf$, x will refer to an up-flex, while it will refer to a down flex on the other side. The case with y is similar. This is consistent notation, since we established that x-down, or x, is equal to x^{-1}, or x-up. Conjugation by r is not trivial, however, and it is straightforward to see that $rxr^3 = y$. Thus the group of motions for this flexagon is given by $(\mathbb{Z}_2 \times \mathbb{Z}_2) \rtimes_\phi D_4$, where $\phi(f)$ is the identity automorphism, and $\phi(r)$ is the automorphism of $\mathbb{Z}_2 \times \mathbb{Z}_2$ sending the generator of the first $\mathbb{Z}_2$, x, to the generator of the second $\mathbb{Z}_2$, y, and vice-versa.

5 A Hamiltonian Cycle Is Not Necessary

Here we encounter an example in which it is possible to perform both an x-down and y-down flex on a single state. This leads to a problem

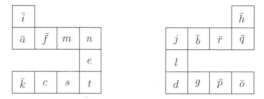

Figure 6.7. Labeled large snake net.

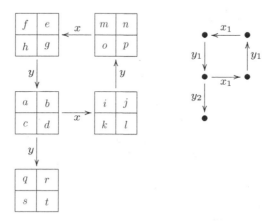

Figure 6.8. Large snake structure diagrams (for the net in Figure 6.7).

mathematizing the flexes. The large snake net is shown in Figure 6.7 and is equivalent to Pook [7, Fig. 6.17]. The description

$$\frac{f}{r} > \frac{q}{a} > \frac{b}{m} \diamond \frac{e}{l} > \frac{\bar{j}}{\bar{n}} \diamond \frac{h}{i} > \frac{\bar{k}}{\bar{o}} \diamond \frac{g}{s} > \frac{t}{d} > \frac{c}{\bar{p}}$$

reveals the folding (tape the bottom of h to $\bar{o}$). This time the structure diagram is not a cycle, as illustrated in the detailed diagram in Figure 6.8. Immediately, this raises questions about group closure and element order. In earlier examples, allowing x and y to correspond to x-down and y-down flexes, respectively, and using the relations $x^2 = y^2 = e$ made sense, because no flexing action could be repeated twice in a row. However, in this flexagon, two y-down flexes can occur in tandem. Yet, notice that the two y-flexes associated with the state $abcd$ function in fundamentally different ways. The y-flex sending $fehg$ to $abcd$ changes the flexagon from a state that is the result of an x-flex to a state that allows an x-flex to be performed. This is the same way that the y on the opposite side of the structure diagram acts. Now the y-flex emanating from $abcd$ takes a state that is the result of a y-flex to a flexagon state that admits only the inverse flex y-up. We call this final state, $qrst$, a

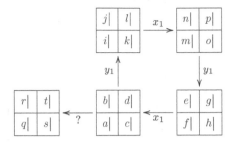

Figure 6.9. Flexes associated with the state $bdac|$.

dead state. These fundamental differences give an organic reason for assigning two different group generators to the two y-down flexes, as in the simplified structure diagram on the right in Figure 6.8.

The group generator x_1 is the x-down flex. Thus we have group generators x_1, y_1, and y_2. With the renaming, no flex may be performed twice in a row, and it again makes sense to require $x_1^2 = y_1^2 = y_2^2 = e$. We easily verify that x_1 and y_1 commute. It is vacuously true that y_1 and y_2 commute, since no state simultaneously allows both $y_1 y_2$ and $y_2 y_1$ to act on it. As before, the next step is to consider the interaction of the dihedral group with these generators. It is clear that f and x_1 commute. Similarly, $f y_i = y_i f$ for $i = 1, 2$. Remember that when y_2 is a down flex on one side of this equation, it will be an up flex on the other side. However, conjugation by r yields some unexpected outcomes. Notice that $r x_1 r^3 = y_1$ and $r y_1 r^3 = x_1$, which is a typical twisting for tetraflexagons. Understanding the element $r y_2 r^3$ is more complicated. Recognize that rotating the state $abcd$ gives the state $bdac|$. From this state, there is only one possible y-flex, as seen in Figure 6.9.

But there is a new x-action that appears after rotating the state $abcd$, labeled with a question mark in Figure 6.9. Since this new x-flex fundamentally results from the rotation of y_2, call this action x_2. It is now easy to verify that $r y_2 r^3 = x_2$. Similarly, conjugation of x_2 by r gives y_2. The previous relations also naturally extend to x_2: $x_2^2 = e$, and x_2 commutes with x_1, y_1, y_2, and f.

At this point the group of actions is known. There are four commuting generators of order two plus the typical dihedral generators. Therefore the group is $(\mathbb{Z}_2 \times \mathbb{Z}_2 \times \mathbb{Z}_2 \times \mathbb{Z}_2) \rtimes_\phi D_4$ where $\phi : D_4 \to Aut(\mathbb{Z}_2 \times \mathbb{Z}_2 \times \mathbb{Z}_2 \times \mathbb{Z}_2)$ is the automorphism defined as follows. The map $\phi(f)$ is the identity, and the map $\phi(r)$ is the automorphism of $\mathbb{Z}_2 \times \mathbb{Z}_2 \times \mathbb{Z}_2 \times \mathbb{Z}_2$ that swaps the generators of two pairs of $\mathbb{Z}_2$s. If one assumes that the generators are x_1, x_2, y_1 and y_2, respectively, then $\phi(r)$ switches the generators of the first and third $\mathbb{Z}_2$ as well as the generators of the second and fourth $\mathbb{Z}_2$.

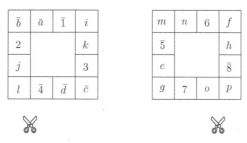

Figure 6.10. Labeled ring net.

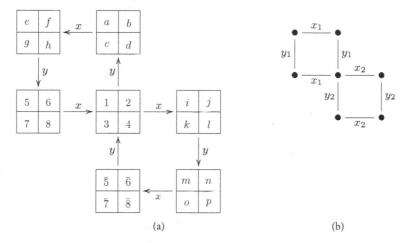

(a) (b)

Figure 6.11. (a) Detailed and (b) simplified structure diagrams for the labeled ring net.

6 Not All Flips Commute

This final example of a flexagon is built from a ring net with the labeling as in Figure 6.10. This example illustrates that the flips in the dihedral group do not always interact nicely with the flex actions. The folding yields what Berkove and Dumont term a crossroad [1, p. 344]. This folding is encoded as

$$\frac{\dot{1}}{n} > \frac{\ddot{m}}{i} \searrow \frac{k}{5} \circ \frac{2}{h} > \frac{\bar{f}}{b} > \frac{a}{6} \circ \frac{3}{e} > \frac{\bar{g}}{c} > \frac{d}{\bar{7}} \circ \frac{4}{\bar{o}} > \frac{\dot{p}}{\bar{l}} > \frac{j}{\bar{8}}$$

(tape the right side of 4 to the square p).

The structure diagram, shown in Figure 6.11, is both closed and an Eulerian cycle, unlike the large snake example of Figure 6.7. The point on the structure diagram that corresponds to the face 1234 is called a *crossroad point*, for obvious reasons.

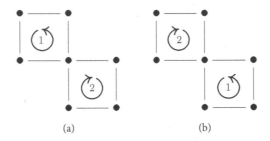

Figure 6.12. Diagrams with crossroad state (a) 1234 and (b) $\bar{6}5\bar{8}\bar{7}$.

Despite the relatively nice structure diagram, an immediate challenge arises. Sometimes two x-down flexes can be performed, and at other times only one x-down flex is possible. Happily, there is again a natural way to split and relabel these xs. We assign group elements x_1 and y_1 to the x- and y-flexes in the upper ring and x_2 and y_2 to those in the lower ring, as shown in Figure 6.11. Consider the two x-flexes tied to the crossroad state. It is clear that one x sends the flexagon into the bottom ring of states and that the inverse of the other transports the flexagon into the upper ring. Further, the movements around these rings are in different directions, as shown in Figure 6.12a. This observation makes our choice to mathematize the actions in this way organic.

Repeating our previous technique, because no flex can be performed repeatedly, we define the relations $x_i^2 = y_i^2 = e$. Additionally, it is easily verified that the flexes commute with each other. The next challenge is to fully understand how the generators of D_4 interact with the x_i and y_i generators. As with the large snake in Section 6, it is easy to verify that conjugation by r exchanges x_i with y_i for $i = 1, 2$. To grasp the action of f, recognize that flipping the flexagon permits all flexes to be performed as before. However, the directions of movement change, which means that the cycle y_1, x_1, y_1, x_1, switches with the cycle y_2, x_2, y_2, x_2. Figure 6.12 illustrates this issue. The left structure diagram is the one obtained with starting state 1234 as the crossroad. The right diagram is obtained by flipping the flexagon and using the state $\bar{6}5\bar{8}\bar{7}$ as the crossroads. The direction of movement is indicated with an internal arrow, and the type of action (1 or 2) is listed in the arrow.

The astute reader will notice that r interacts with the flexes here in the same way as it did in the previous example. However, unlike in all of our previous cases, f does not commute with the flex actions. We now state the group explicitly. The dihedral group D_4 twists the four commuting generators of order two, which we list in the order x_1, y_1, x_2, y_2. So the action group is $(\mathbb{Z}_2 \times \mathbb{Z}_2 \times \mathbb{Z}_2 \times \mathbb{Z}_2) \rtimes_\phi D_4$, where $\phi : D_4 \to Aut(\mathbb{Z}_2 \times \mathbb{Z}_2 \times \mathbb{Z}_2 \times \mathbb{Z}_2)$ is the automorphism defined as follows. The automorphism $\phi(f)$ is $\phi(f)(x_i) = x_j$ and $\phi(f)(y_i) = y_j$ for $1 \leq i, j \leq 2$, $i \neq j$. The automorphism $\phi(r)$ is $\phi(r)(x_i) = y_i$ and $\phi(r)(y_i) = x_i$ for $i = 1, 2$.

7 Conclusion

After joining the notions of symmetry group and flexagon group, and being careful about the mathematizing of flexes, interesting groups are admitted by many flexagons. The process of finding a group of actions involves many choices about how to interpret the physical world, but here all choices were made for natural reasons. Contrary to the previous hypothesis in the literature, a single closed cycle is not a requirement for a group structure. Furthermore, neither a Hamiltonian nor Eulerian cycle is required, though many of the examples presented in this chapter have such a cycle.

While progress has certainly been made with regard to groups and tetraflexagons, many questions remain. Here x- and y-flexes are frequently split into two types. Is it possible to find groups without splitting the flexes by lifting the flexagon to a higher dimension or embedding it inside of a larger set of objects? Are these flexagon actions simply the projection of a larger space in different ways? If so, what is being projected, and what do we learn? The similarity of the groups in the final two examples, the large snake and the ring nets, respectively, is striking. These examples have different nets and different structure diagrams. What are the physical invariants that would be useful in determining the group structure? These and other serious mathematical questions about innocent looking tetraflexagons await greater exploration.

References

[1] E. Berkove and J. Dumont. It's okay to be square if you're a flexagon. *Math. Mag.* **77** no. 5 (2004) 335–348.

[2] M. Gardner. Flexagons. *Sci. Am.* **195** (June 1956) 162–168.

[3] V. Hart. *Hexaflexagons.* https://www.youtube.com/watch?v=VIVIegSt81k (accessed August 24, 2014).

[4] P. Hilton, J. Pedersen, and H. Walser. Faces of the tri-hexaflexagon. *Math. Mag.* **70** no. 4 (1997) 243–251.

[5] T. Hungerford. *Algebra. Graduate Texts in Mathematics* **73**. Springer-Verlag, New York, 1974.

[6] J. Pedersen. Sneaking up on a group. *Two-Year Coll. Math. J.* **3** no. 2 (1972) 9–12.

[7] L. Pook. *Flexagons Inside Out.* Cambridge University Press, Cambridge, 2003.

7

PARALLEL WEIGHINGS OF COINS

Tanya Khovanova

I heard my first coin problem when I was very young:

Given nine coins, one of them fake and lighter, find the fake coin in two weighings on a balance scale.

I believed that this problem was thousands of years old, that Pythagoras could have invented it. But, surprisingly, its first publication was by E. D. Schell in the January 1945 issue of the *American Mathematical Monthly* [9]. This would seem to make it most unlikely that Pythagoras knew of the problem.

This is a well-known problem, but it is useful for establishing the methods that are used later, in my featured problem. The problem implies that all real coins weigh the same, and that we need a strategy that guarantees finding the fake coin in two weighings. Two weighings is the smallest number that guarantees finding the fake coin. In addition, nine is the largest number of coins such that the fake can be found in two weighings. Rather than examining the solution for the nine-coin problem, I shall discuss a well-known generalization to any number of coins:

Given N coins, one of them fake and lighter, find the minimum number of weighings that will guarantee finding the fake coin.

Here is a way to think about it: In one weighing, divide the coins into those that go onto the left pan, those that go onto the right pan, and those that do not go on the scale. Clearly, we need to put the same number of coins on the pans; otherwise, we do not get any meaningful information. If the scale balances, then the fake coin is in the leftover pile. If the scale does not balance, then the fake is in the pile that is lighter. Either way, the coin is in one of the three piles, and we can start over with this smaller pile. So, to minimize the number of weighings, we divide all the coins into three piles so as to minimize the size of the largest pile. Dividing the coins into three piles as evenly as possible allows us, with n weighings, to find the fake coin from among up to 3^n coins. In particular, we can find the fake among nine coins in two weighings.

Another famous coin puzzle appeared around the same time as the previous puzzle [3]:

There are twelve coins; one of them is fake. All real coins weigh the same. The fake coin is either lighter or heavier than the real coins. Find the fake coin, and decide whether it is heavier or lighter in three weighings on a balance scale.

The solution is well known and quite beautiful. Unsurprisingly, it generated more publications than the nine-coin problem [1, 5, 6, 11]. Readers who do not know the solution should try it.

What is the minimum number of weighings in this puzzle's setting for an arbitrary number of coins? Note that we need to assume that the number of coins is more than two; otherwise we cannot resolve it at all.

If there are N coins, then there are $2N$ possible answers to this puzzle. We need to pinpoint the fake and say whether it is heavier or lighter. Each weighing divides information into three parts, so in n weighings we can give 3^n different answers. Thus, the expected number of weighings should be of the order $\log_3 2N$. The exact answer can be calculated using the additional constraint of having the same number of coins on each pan in each weighing. The exact answer is $\lceil \log_3(2N + 3) \rceil$ (see [2, 4]). Equivalently, the maximum number of coins for which the problem can be solved in n weighings is $(3^n - 3)/2$.

In the following important variation of the latter puzzle, we need to find the fake coin, but we do not need to tell whether is it heavier or lighter [2].

There are N coins; one of them is fake. All real coins weigh the same. The fake coin is either lighter or heavier than the real coins. What is the maximum number of coins for which you can guarantee finding the fake coin with n weighings on a balance scale?

This problem is similar to the previous one. Let me call the previous problem the *find-and-label* problem, and I will call this problem the *just-find* problem.

The answer to the just-find problem is $(3^n - 1)/2$ (see [8]). In particular, thirteen coins is the best we can do using three weighings.

Notice that for every strategy for the find-and-label problem that resolves n coins, we can produce a strategy for the just-find problem that resolves $n + 1$ coins by adding a coin that is never on the scale. Indeed, with the find-and-label strategy, by the last weighing, at least one of the weighings needs to be unbalanced to label the fake coin. Thus, if all the weighings balance at the end, the fake is the extra coin.

Let us now change our focus to more modern coin-weighing puzzles. We have all been hearing about parallel computing, and now it is used in a puzzle invented by Konstantin Knop. The puzzle appeared at the 2012 Ukraine-Russia Puzzle Tournament [10] and in Knop's blog [7].

We have N indistinguishable coins. One of them is fake, and it is not known whether it is heavier or lighter than the genuine coins, which all weigh the same. There are two balance scales that can be used in parallel. Each weighing lasts one minute. What is the largest number of coins N for which it is possible to find the fake coin in five minutes?

In the remainder of this chapter, I present my solutions to generalizations of Knop's puzzle for any number of minutes and any number of parallel scales. Section 1 describes the similarity of the original puzzle with a multiple-pans problem: a coin-weighing puzzle involving balance scales with not two, but any number of pans. The notion of a coin's potential—a useful technical tool in solving coin-weighing puzzles—is defined in Section 2. The number of coins with known potential that can be processed in n minutes is discussed there too. Section 3 provides a solution to the parallel weighing problem in case we have an unlimited supply of real coins. It is followed by a solution to the original puzzle and its generalization for any number of minutes in Section 4. Section 5 generalizes these results to the use of more than two scales in parallel. The find-and-label variation of this problem for any number of minutes is discussed in Section 6. Finally, Section 7 compares the find-and-label problem with the just-find problem.

1 Warm-up: The Multiple Pans Problem

Knop's puzzle is reminiscent of another coin-weighing problem, where in a similar situation we need to find a fake coin by using five weighings on one scale with four pans. The answer in this variation is $5^5 = 3,125$. Divide the coins into five groups, each with the same number of coins, and put four groups on the four pans of the scale. If one of the pans is different (heavier or lighter), then this pan contains the fake coin. For example, a heavier coin on a pan leads to that pan being lower than the other three pans. Thus, after the weighing, we know whether the fake coin is heavier or lighter than the real coins. When the four pans balance, the leftover group contains the fake coin. The strategy is to divide the coins into five piles as evenly as possible. This way, each weighing reduces the pile with the fake coin by a factor of five. Thus it is possible to resolve 5^n coins in n weighings.

I leave it to the reader to check that, excluding the case of two coins, any number of coins greater than 5^{n-1} and not greater than 5^n can be optimally resolved in n weighings.

One scale with four pans gives more information than two scales with two pans used in parallel. Therefore, Knop's puzzle requires at least the same number of weighings as the four-pan puzzle for the same number of coins.

So the answer to Knop's puzzle, using a pair of scales and five weighings each, does not exceed 3,125. But what is it?

2 A Coin's Potential

While weighing coins, we may be able to determine some incomplete information about its authenticity. For instance, we may be able to rule out the possibility that a given coin is fake-and-heavy, without being able to tell whether that coin is real or fake-and-light. Let us call such a coin *potentially light*; and conversely, let us say a coin is *potentially heavy* if it could be real or fake-and-heavy but cannot be fake-and-light.

How many coins with known potential can be processed in n minutes?

If all the coins are potentially light, then we can find the fake coin out of 5^n coins in n minutes. Indeed, in this case, using two scales or one scale with four pans (see Section 1) gives us the same information. The pan that is lighter contains the fake coin. If everything balances, then the fake coin is not on the scales.

What if there is a mixture of potentials and we use two, two-pan scales? Can we expect the same answer? How much more complicated could it be? Suppose there are five coins: two of them are potentially light, and three are potentially heavy. Then on the first scale, we compare one potentially light coin with the other such coin. On the other scale, we compare one potentially heavy coin against another potentially heavy coin. The fake coin can be determined in one minute.

Our intuition suggests that it is a bad idea to compare a potentially heavy coin on one pan with a potentially light coin on the other pan. Such a weighing, if unbalanced, will not produce any new information. Instead, if we compare a potentially heavy coin with a potentially heavy coin, then we will get new information. If the scale balances, then both coins are real. If the scale does not balance, then the fake coin is the heavier coin of the two that are potentially heavy.

Does this mean that we should only put coins with the same potential on the same scale? Actually, we can mix the coins. For example, suppose we put three potentially light coins and five potentially heavy coins on each pan of the same scale. If the left pan is lighter, then the potentially heavy coins on the left pan and potentially light coins on the right pan must be genuine. The fake coin must be either one of the three potentially light coins on the left pan or one of the five potentially heavy coins on the right pan.

In general, after each minute, the best hope is that the number of coins that are not determined to be real is reduced by a factor of five. If one of the weighings on one scale is unbalanced, then the potentially light coins on the lighter pan, plus the potentially heavy coins on the heavier pan must contain the fake coin. We do not want this number to be bigger than one-fifth of the total number of coins being processed. So, take coins in pairs having the same potential, and from each pair put the coins on different pans of the same scale.

In one minute we can divide the group into five equal, or almost equal, groups. If there is an odd number of coins with the same potential, then the extra coin does not go on the scales.

The only thing left to check is what happens when the number of coins is small. Namely, we need to check what happens when the number of potentially light coins is odd, the number of potentially heavy coins is odd, and the total number of coins is not more than five. In this case the algorithm requires us to put aside two coins: one potentially heavy and one potentially light, but the put-aside pile cannot have more than one coin.

After checking small cases, we see that we cannot resolve the problem in one minute when there are two coins of different potential, or when the four coins are distributed as one and three. But if extra coins are available that are known to be real, then these two cases can be resolved. So the small cases are a problem only if they happen in the first minute. Thus we have established the following.

Lemma 1. *Any number of coins $N > 4$ with known potential can be resolved in $\lceil \log_5 N \rceil$ minutes.*

3 Unlimited Supply of Real Coins

We say that a coin that is potentially light or potentially heavy has *known potential*. The following theorem shows why the notion of known potential is important in solving Knop's puzzle and in many other coin-weighing puzzles as well.

Theorem 1. *In a coin-weighing puzzle where only one coin is fake, any coin that visited the scales is either genuine or its potential is known.*

Proof. If a scale balanced, all coins that were on it are real. Any coin that appeared on both a heavier pan and a lighter pan is also real. Otherwise, the coins that only visited lighter pans are potentially light, and the coins that only visited heavier pans are potentially heavy. □

Now let us go back to the original problem, in which initially we do not know the coins' potential. Let us temporarily add an additional assumption. Suppose there is an unlimited supply of coins that we know are real. Let $u(n)$ be the maximum number of coins we can process in n minutes if we do not know their potential but have an unlimited supply of real coins.

Lemma 2. *The largest number of coins we can process in n minutes is*

$$u(n) = 2 \cdot 5^{n-1} + u(n - 1).$$

Proof. What information do we get after the first minute? Both scales might be balanced, meaning that the fake coin is in the leftover pile of coins with unknown potential. So we then have to leave out at most $u(n - 1)$ coins. Otherwise, exactly one scale is unbalanced. In this case, all the coins on this scale will have their potential revealed. The number of these coins cannot be more than 5^{n-1}, so

$$u(n) \leq 2 \cdot 5^{n-1} + u(n - 1).$$

Can we achieve this bound? Yes. On each scale, put 5^{n-1} unknown coins on one pan and 5^{n-1} real coins from the supply on the other. Thus

$$u(n) = 2 \cdot 5^{n-1} + u(n - 1).$$

$\square$

We also can see that $u(1) = 3$. Indeed, on each scale put one coin against one real coin and have one coin in the leftover pile. The following corollary now follows easily using induction.

Corollary 1. $u(n) = (5^n + 1)/2.$

The answer to the puzzle problem with the additional resource of an unlimited supply of real coins is therefore $(5^n + 1)/2$. The answer without the additional resource clearly cannot be larger. But what is it?

We assumed that there is an unlimited supply of real coins. But how many extra coins do we really need? The extra coins are needed for the first minute only, because after the first minute, at least one of the scales will balance, and many coins will be determined to be real. In the first minute, we need to put 5^{n-1} coins from the unknown pile on each scale. The coins do not have to be on the same pan. The difficulty is that the number of coins is odd, so we need one extra real coin to make this number even. Thus our supply need not be unlimited—we need only two extra coins, one for each scale.

4 The Solution

Recall that the formula for $u(n)$ assumes an unlimited supply of real coins. But we have seen that the unlimited supply need not be more than two additional real coins.

So, how can we solve the original problem? We know that the only adjustment needed is in the first minute. In that minute we put unknown coins against unknown coins, not more than 5^{n-1} on each scale. Since the number on each scale must be even, the best we can do is put $5^{n-1} - 1$ coins on each. Thus, the answer to the puzzle is $(5^n - 3)/2$.

Of course we cannot ever find the fake coin out of two coins.

Theorem 2. *Given two parallel scales, the number of coins N that can be optimally resolved in exactly n minutes is*

$$\frac{5^{n-1} - 3}{2} \le N < \frac{5^n - 3}{2},$$

with one exception: $N = 2$, for which the fake coin cannot be identified.

In Knop's original puzzle, the largest number of coins that can be resolved in five minutes is 1,561.

5 More Scales

It is straightforward to generalize the just-find problem to any number of scales used in parallel. Suppose the number of scales is k. The following problems can be resolved in n minutes.

Known Potential. If all the coins have known potential, then any number of coins up to $(2k + 1)^n$ can be resolved.

Unlimited Supply of Real Coins. If we do not know the potential of any coin and there is an unlimited supply of real coins, the maximum number of coins that can be resolved is defined by a recursion: $u_k(n) = k \cdot (2k + 1)^{n-1} + u_k(n - 1)$ and $u_k(1) = k + 1$. Any number of coins up to $((2k + 1)^n + 1)/2$ can be resolved.

General Case. If we do not know the potential of any coin and there are no extra real coins, then any number of coins between 3 and $u_k(n) - k = ((2k + 1)^n + 1)/2 - k$ can be resolved.

Note that if $k = 1$, then the general case is the classic problem of just finding the fake coin. So plugging $k = 1$ into the formula $((2k + 1)^n + 1)/2 - k$ above should give the answer provided in Section 1: $(3^n + 1)/2 - 1$. In particular, for $n = 3$ it should be 13. And it is!

6 Find and Label

The methods described above can be used to solve another problem in the same setting: Find the fake coin and say whether it is heavier or lighter. When all coins have known potential, the just-find problem is equivalent to the find-and-label problem.

The find-and-label problem can be resolved by similar methods to the just-find problem. Namely, let us denote by $U_k(n)$ the number of coins that can be resolved in n minutes on k parallel scales when there is an unlimited supply of extra real coins. Then the recursion is the same as for the just-find problem: $U_k(n) = k \cdot (2k+1)^{n-1} + U_k(n-1)$. The difference from the just-find problem is in the starting point: $U_k(1) = k$.

If we do not have an unlimited supply of real coins, the bound is described by the following theorem.

Theorem 3. *Suppose there are N coins, one of which is fake, and it is not known whether it is heavier or lighter than the real coins. Assume there are k balance scales that can be used in parallel, one weighing per one minute. The maximum number of coins for which we can find and label the fake coin in n weighings is $((2k + 1)^n + 1)/2 - k - 1$. If $N = 2$, then the problem cannot be resolved.*

If $k = 1$, then this is the classic problem of just funding and labeling the fake coin. So plugging $k = 1$ into the formula $((2k + 1)^n + 1)/2 - k - 1$ above should give the answer from Section 1: $(3^n + 1)/2 - 2$. In particular, for $n = 3$, we get 12.

7 Lazy Coin

You might have noticed that the answers for the just-find and for the find-and-label problems differ by one.

Lemma 3. *In the parallel weighing problem with one fake coin, the maximum number of coins that can be optimally resolved for the just-find problem is one more than the maximum number of coins that can be optimally resolved in the find-and-label problem using n weighings.*

We have already proved the lemma by calculating the answer explicitly. It would be nice if there was a simple argument to prove it without calculations. Such an argument exists if we restrict ourselves to static strategies. In a *static* or *nonadaptive* strategy, we decide beforehand what our weighings are. Then, using the results of the weighings, we can find the fake coin and label it if needed.

In a static strategy for the find-and-label problem, every coin has to visit the scales at some point. Otherwise, if the coin does not visit the scales and it happens to be fake, it cannot be labeled. In a static strategy for the just-find problem, if every coin visits the scale, then all the coins can be labeled.

Suppose we add an extra coin that is never on the scales. If all the weighings balance, then the extra coin is the fake one. We cannot have two such coins. Indeed, if one of them is fake, then we cannot differentiate between them. If all the coins visit the scales in the just-find problem, then all the coins can be labeled eventually.

Let us call a strategy that resolves the maximum number of coins in a given number of weighings a *maximal* strategy. We just showed that a maximal static strategy for the just-find problem has to have a coin that does not go on the scales. Therefore, a bijection between maximal static strategies exists for the just-find and the find-and-label problems. The strategies differ by an extra coin that always sits lazily outside the scales.

In *dynamic* or *adaptive* strategies, the next weighing depends on the results of the previous weighings. With dynamic strategies, the story is more complicated. There is no bijection in this case.

It is therefore possible to add a lazy coin to a strategy in the find-and-label problem to get a strategy in the just-find problem. But maximal strategies exist in the just-find problem in which all the coins can end up on the scale.

For example, consider the following strategy to just-find the fake coin out of four coins in two weighings on one scale. In the first weighing, we balance the first coin against the second. If the weighing unbalances, we know that one of the participating coins is fake, and we know the potential of every participating coin. In the second weighing, we balance the first and second coins against the third and fourth. In this example, all coins might visit the scale, and four is the maximum number of coins that can be processed in two weighings.

Alas! There is no simple argument, but at least it is easy to remember that the maximal strategies for these two problems differ by one coin.

Acknowledgments

I am grateful to Daniel Klain and Alexey Radul for helpful discussions. I am also grateful to an enthusiastic anonymous reviewer for thoroughly reading the paper and providing many comments.

References

[1] B. Descartes. The twelve coin problem. *Eureka* **13** (1950) 7, 20.
[2] F. J. Dyson. Note 1931—The problem of the pennies. *Math. Gaz.* **30** (1946) 231–234.
[3] D. Eves. Problem E712—The extended coin problem. Am. Math. Monthly **53** (1946) 156.
[4] N. J. Fine. Problem 4203—The generalized coin problem. Am. Math. Monthly **53** (1946) 278. Solution, 54 (1947) 489–491.

[5] R. L. Goodstein. Note 1845—Find the penny. *Math. Gaz.* **29** (1945) 227–229. [Erroneous solution]

[6] H. D. Grossman. The twelve-coin problem. Scripta Math. **11** (1945) 360–361.

[7] K. Knop. Weighings on two scales. http://blog.kknop.com/2013/04/blog-post_11.html. 2013 (in Russian).

[8] J. G. Mauldon. Strong solutions for the counterfeit coin problem. IBM Research Report RC 7476 (#31437), 1978.

[9] E. D. Schell. Problem E651—Weighed and found wanting. *Am. Math. Monthly* **52** (1945) 42.

[10] 2012 Ukraine-Russian Puzzle Tournament. http://kig.tvpark.ua/_ARC/2012/KG_12_38_12.PDF. 2012. (in Russian)

[11] Lothrop Withington. Another solution of the 12-coin problem. *Scripta Math.* **11** (1945) 361–363.

8

ANALYSIS OF CROSSWORD PUZZLE DIFFICULTY USING A RANDOM GRAPH PROCESS

John K. McSweeney

One of the most common recreational pursuits in North America is the crossword puzzle. What distinguishes a crossword puzzle from a simple list of trivia questions is that the answers are entered into a grid in crossing fashion, and therefore each correct answer obtained provides partial information about others. This interdependence among answers affects the relationship between the individual clue difficulties and the final number of answers that can be obtained. Indeed, even if there are only a few easy answers that can be found immediately, these may trigger further answers, and, in such a cascading fashion, many or all of the answers in the puzzle may be found.

In this chapter, I quantify these dynamics by using a network structure to model the puzzle. Specifically, I determine how the interaction between the structure of cells in the puzzle and the difficulty of the clues affects the puzzle's solvability. This is achieved by viewing the puzzle as a network (graph) object in which answers represent nodes, and answer crossings represent edges. I assign a random distribution of difficulty levels to the clues and model the solution process with the so-called *cascade process* developed by Watts [11]. This randomness in the clue difficulties is ultimately responsible for the wide variability in the solvability of a puzzle, which many solvers know well—a solver, presented with two puzzles of ostensibly equal difficulty, may solve one readily and be stumped by the other.

The structure of the chapter is as follows. First, for a simple class of highly symmetric networks, I build an iterative stochastic process that exactly describes the solution and obtain its deterministic approximation, which gives a very simple fixed-point equation to solve for the final solution proportion. I then show via simulation on actual crosswords from the Sunday edition of *The New York Times* that certain network properties inherent to actual crossword networks are important predictors of the final solution size of the puzzle. I also include, in Appendix I, visualizations of how the mean and standard deviation of the clue difficulties affect the total proportion of answers that can be found. Appendix II shows some results of partial solutions on actual crossword grids.

1 Preliminaries

This section outlines the basic properties and definitions relating to standard English-language North American crossword puzzles. There are certainly puzzles in other languages or with extra gimmicks that do not obey all the following rules, but we won't be concerned with them (except for cryptic crosswords, discussed in Section 3.1).

A crossword puzzle consists of a square $k \times k$ grid made up of white and black squares; typical values of k are fifteen for weekday *New York Times* puzzles and twenty-one for their famous Sunday puzzles. White squares will also be called *cells*; exactly one letter belongs in each white square, and nothing goes into the black squares. An uninterrupted horizontal or vertical sequence of white squares, beginning and ending at a black square or at the puzzle's edge, is called an *answer*; there are two types of answers, *Across* (A) for horizontal and *Down* (D) for vertical. An answer is often a single English word, but it could also be a name, acronym, phrase, or word-part (e.g., suffix), for example. By convention, each answer in a puzzle must be of length at least three; define $\ell(x)$ to be the length (number of letters) in an answer x.

Each answer has its own clue, though often the clue alone is not sufficient to find the answer. For example, if the clue is "One of the Beatles," and the answer has length four, then there is no way to determine whether the answer is JOHN or PAUL without first finding crossing answers. If an answer is found, then all of its letters are entered into the grid. There is a standard way to label the answers and clues, by starting at the top left of the grid, scanning the grid row-by-row, and sequentially entering $1, 2, 3, \ldots$ into any cell that starts an answer (Across, Down, or both); however, we will not be concerned with the precise way in which this numbering is done. Note that every letter (cell) is part of exactly two answers, one Across and one Down. Typically, there are more white than black squares, although it is not necessary to assume this for the subsequent analysis. A rule of thumb is that no more than one-sixth of the squares should be black.

The final *solution set* is the set of all answers that the solver is able to enter into the grid. The puzzle is fully solved if every letter is successfully entered (equivalently, every answer is found). The solvability of a puzzle depends, broadly, on three criteria: the arrangement of answers in the grid; the inherent difficulty of the clues; and the skill and/or knowledge base of the solver. These last two are clearly interdependent and subjective. When talking about the *difficulty* of a clue or a puzzle, it should be understood that this is only with respect to a particular solver. The same puzzle with a different solver (with a

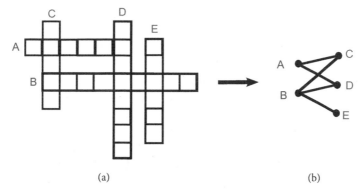

Figure 8.1. The crossword puzzle (a) is translated to a network (b) drawn in the familiar node/edge graphical form. Answers A and B are Across; answers C,D, and E are Down. Notice the lack of an edge between A and E in the network, since answers A and E do not cross.

different knowledge base, eye for patterns, etc.) would experience a different degree of difficulty. For what follows, it may be helpful to think of some fixed universal solver, so that the level of difficulty varies across clues and puzzle structures.

2 The Puzzle as a Network

To quantify the dynamics of a solution, a natural choice is to model the puzzle with a network (or graph) structure: a set V of nodes (or vertices), and a set E of edges (or links) between certain pairs of them. In this analysis, we will not assign weights or directions to the edges. Each answer in the puzzle is identified with a node in the network, and we put an edge between two nodes x and y in the network if the corresponding answers in the puzzle cross. The *degree* of a node is defined to be the number of edges incident to it. In a standard crossword, every cell is part of two words. This implies that the degree of a node is equal to the length of the word. (Contrasts this with the case of a cryptic crossword, as shown in Figure 8.3 in Section 4.) The network so created from a puzzle is called the puzzle's *crossword network*. The puzzle fragment in Figure 8.1 gives a simple example of how this network is created.

The following graph properties are characteristic of networks generated from crossword puzzles:

- *Bipartiteness.* This property follows from the definition: no two answers of the same orientation (Across or Down) may cross, so the Across/Down division yields a bipartition $V = A \cup D$. A and D are not necessarily of the same size, but the conservation relation

$$\sum_{x \in A} \ell(x) = \sum_{x \in D} \ell(x) = \text{ total number of cells in the puzzle}$$

always holds.

- *Degree Variability*. There is often a small set of "theme" answers that are much longer than the rest of the answers in the puzzle, hence the crossword network has nodes with widely varying degrees.
- *Modularity*. This property is difficult to quantify precisely. Broadly, a network is defined to be highly *modular* if its node-set can be partitioned into blocks in such a way that there are many edges in each block of the partition and few between the blocks. In general network theory, the problem of quantifying this notion, and of finding a modular partition of a given network, is an active area of research; it is referred to as *community detection* in the social science literature (e.g., [3]). For the case of crosswords, however, it is often clear how such a partitioning can reasonably be done—blocks of nodes in a modular partition of a puzzle's network correspond to large uninterrupted clusters of white squares in the puzzle grid.
- *Clustering*. Crossword networks have many short cycles; for instance, any 2×2 block of cells in the puzzle yields a cycle of length 4 (a "square") in the puzzle network. The high number of these squares creates cyclical dependencies among the statuses of the answers and may produce a qualitative effect on the dynamics. For instance, suppose x_1 and x_2 are Across answers that each cross two Down answers y_1 and y_2. Solving x_1 may not be directly sufficient to allow y_1 to be found, but it may do so indirectly, via a cascade of answers $x_1 \rightarrow y_2 \rightarrow x_2 \rightarrow y_1$. A common measure of the degree of clustering of a network is the clustering coefficient (see, e.g., Watts and Strogatz [12]) which involves a census of the number of triangles (i.e., induced C_3 subnetworks) in the network. This number is, of course, no use to us, since it is trivially zero for the bipartite networks considered here. I therefore propose in Section 5.1.1 an alternative definition of the clustering coefficient for bipartite networks.

3 Difficulty Thresholds and Solution Process

Clearly, an answer becomes easier to find, and certainly no harder, once some of its letters are determined (by finding crossing answers). To model this mathematically, we will associate, to each clue x, a *difficulty threshold* $\varphi_x \in \mathbf{R}$. and let us posit the following rule.

Update Rule. *An answer x can be found if and only if the proportion of its letters that have been found so far is greater than or equal to φ_x.*

Example. To demonstrate why it is a sensible modeling assumption to consider the proportion of letters present (and not, for instance, the absolute number), consider the following three possibilities for a partially found answer:

1. C___ ___
2. C___ ___ ___ ___ ___ ___ ___ ___ ___ ___
3. CATERPILL___R

Clearly, all else being equal, the second of these is the most difficult to complete (1/11 letters present), the first is next most difficult (1/3 letters), and the third is the easiest (10/11 letters). To see how the difficulty threshold φ enters into it, let's focus on the second word. If the clue for the second is "A Kind of Animal," then the solver might need half the letters to be present to deduce the full answer (CATERPILLAR), which would correspond to a clue with difficulty threshold around $\varphi = 1/2$. However, if the clue were the easier "Heavy Equipment Manufacturer Based in Peoria, Ill.," then perhaps the solver would only need the one letter to be sure of the correct answer, which would correspond to a difficulty level of $\varphi \leq 1/11$.

An equivalent way to state the update rule is that an answer x can be found once the proportion of its crossing answers that are found is at least φ_x. (Usually the φ_x will take values in $[0, 1]$, but it will be convenient to allow them to take other values as well, as explained below.) Note that we do not allow an answer to be partially solved on its own: every letter found is part of at least one fully solved answer. This restriction precludes situations where the solver may find only one part of a compound word or phrase, or may know the right answer but be unsure of its spelling.

With this formalism, answers x with $\varphi_x \leq 0$ can be found 'spontaneously', and serve to seed the grid with letters. Clearly, if there are no answers with threshold ≤ 0, then no answers can ever be found (unless the grid is seeded with answers in some artificial fashion). To borrow terminology from epidemiology, answers x with $\varphi_x \leq 0$ will be called *seeds;* a seed in this context is the analogue of a "Patient Zero" in the spread of an epidemic. An answer x with $\varphi_x > 1 - (1/\ell(x))$ cannot be found until all its letters are found; such answers are called *resistant.* By convention, an answer x is solved if all of its letters are found, even if its difficulty threshold φ_x is greater than 1. Note that if two resistant answers cross, the letter in the cell where they cross can never be found.

Example. Consider the puzzle fragment in Figure 8.2, and suppose that the difficulty thresholds for the answers are as follows:

$$\varphi_A = -0.03, \quad \varphi_B = 0.10, \quad \varphi_C = 0.38, \quad \varphi_D = 0.08, \quad \text{and} \quad \varphi_E = 0.21.$$

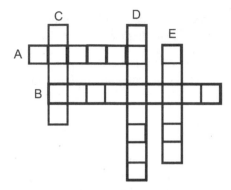

Figure 8.2. Puzzle fragment with difficulty thresholds $\varphi_A = -0.03$, $\varphi_B = 0.10$, $\varphi_C = 0.38$, $\varphi_D = 0.08$, *and* $\varphi_E = 0.21$. All answers except E can be found with these thresholds.

Then we have the following dynamics

- Answer A can be found spontaneously, since $\varphi_A \leq 0$.
- Answer D can then be found, because the proportion of its letters that are found is $1/8$, which is greater than its threshold $\varphi_D = 0.08$. Note that answer C cannot (yet) be found, since $\varphi_C = 0.38 > 1/5$.
- Answer B has can now be found from the letter provided by D, since $\varphi_B = 0.10 < 1/9$.
- Answer C can now be found, since two of its letters have been found, and $\varphi_C = 0.38 < 2/5$.
- Answer E can never be found in this puzzle fragment, since $\varphi_E = 0.21 > 1/6$, and there are no more words that cross E.

3.1 Cryptic Crosswords

Some puzzle properties discussed in Section 1 do not apply to so-called *cryptic crosswords*. In particular, there are many cells that belong to only one answer, and thus the grid is a good deal sparser than for ordinary crossword puzzles (see Figure 8.3). It is clear then that to admit a full (or nearly full) solution, the difficulty level of clues in a cryptic crossword must be much lower on average than those in a standard puzzle. Indeed, Figure 8.7 in Appendix 1 demonstrates this clearly. Note that by "difficulty," we mean not the amount of time or effort needed to solve a clue, but rather a binary value (a function of how many crossing answers have been found) of whether or not it can be solved, regardless of the time or ingenuity that this may take. Indeed, many cryptic clues may be figured out without the help of crossing letters but with a good deal of reflection.

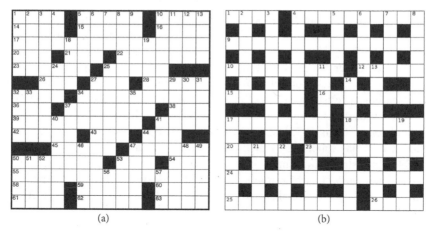

Figure 8.3. (a) Standard crossword puzzle grid; (b) cryptic crossword puzzle grid. In the cryptic puzzle, note the presence of cells that belong to only one answer.

3.2 Relation to Network-Based Epidemic Modeling

There has been much recent work on epidemic-type processes on networks, in which each node has a (time-evolving) state, such as Susceptible (S), Infective (I), Exposed (E), or Recovered (R). States of nodes are affected by the current states of their neighboring nodes, in a potentially random fashion. Examples include SIS, SIR, and SI disease models [5, 6], or bond, site, and bootstrap percolation [1]. A fundamental question for such processes is the following. Starting with a small number of seed nodes in some "infected" or "active" (or "solved") state, what features of the network and of the disease-spreading mechanism determine whether it is possible (or how likely it is) for the disease to propagate—that is, to activate or infect a large fraction, or the totality, of the population? In the context of a crossword network, consider an answer to be active if it is solved. By using the *proportion* of active neighbors as the criterion for activation, as done here, we have an instance of what is known as *Watts' Cascade Process* [11]. In purely network-theoretic terms, this process is defined succinctly as follows. For a network $G = (V, E)$, all nodes start off inactive, and a function $\varphi : V \to \mathbf{R}$ is given. Each node x is activated if and only if the proportion of its neighbors that are active reaches at least φ_x, and once a node is active, it remains so forever.

3.3 Random Thresholds

If all the answers had some identical difficulty threshold $\varphi > 0$, then the solution can never get started, unless we planted seeds in the puzzle in some artificial fashion. Therefore, as a natural way to introduce variability in clue difficulty, we take the thresholds $\{\varphi_x\}$ to be independent and identically distributed (iid) random variables with common cumulative distribution function (cdf) F. Like any recreational endeavor, a good crossword puzzle should be neither too easy nor too difficult. Therefore, an appropriate puzzle should be completely solvable, save possibly for a few isolated letters at the intersection of resistant answers; it should also not be too easy, in that there should be a fairly large number of clues of satisfyingly challenging intermediate difficulties. A Normal (Gaussian) distribution for the $\{\varphi_x\}$ seems like a natural choice, as it has small tails, which create some, but not many, seed answers to get things started and make for a pleasing cascading solution process, and few resistant answers, which might make the puzzle unsolvable. A good puzzle will have as large as possible a mean difficulty level while still being fully solvable.

3.4 Limitations of the Model

Like any mathematical model, this process makes many unrealistic simplifying assumptions, and potentially important features of the output are ignored; let us review these issues here briefly. First, some letters may be more helpful than others with regard to finding an answer; for example, if a crossing Q is found, then the next letter is likely to be U, and moreover, there will be few potential answers that have a Q in that position. Indeed, adjacent letter pairs are far from being evenly distributed (e.g., consonants and vowels tend to alternate). Also, it is often possible to find part of an answer but not the whole thing, especially if the answer is actually a phrase that consists of several words.

In addition, a puzzle is often based on a theme, so that certain long answers have a common topic or employ a common linguistic trick. In this case, solving one theme answer may elucidate the trick in question, thereby making the other theme answers easier to find, even though they may not directly cross. This could be modeled by introducing extra edges in the puzzle network to represent the thematic connection between these theme words.

3.5 Structure of the Chapter

A network-based process such as the one presented in this chapter is inherently high-dimensional: if there are n answers in the puzzle, then there are n difficulty thresholds to generate, and 2^n possible states of the system, since each answer may be solved or not. For this reason, exact analytical answers are difficult or impossible to come by, except in the case of networks that are very small or have a high degree of symmetry. Our analysis will therefore proceed as follows. Section 4 considers the case where the puzzle is a fully white $m \times m$ grid (and hence whose network is a complete bipartite network $K_{m,m}$). In that case, we can afford to keep track only of the *number* of solved answers in each direction (and not their *locations*), rendering the problem two-dimensional. We derive an (exact) Markov-like stochastic process to describe the evolution of its solution and then analyze the deterministic process that arises as a limiting case in the large-graph limit ($m \to \infty$). Section 5 builds networks from actual crossword puzzles taken from the Sunday *New York Times* and presents some (low-dimensional) summary network statistics that are likely to be relevant to the solution process. We then examine the results of simulations of the process (with R) across a variety of threshold distributions and determine the relationship between the moments (mean and standard deviation) of the threshold distribution and the solution set of the puzzle.

4 Completely White Grid

Consider the highly symmetric case where the puzzle grid is an $m \times m$ full white square, which means that the associated crossword network is a complete, bipartite graph $K_{m,m}$, since every Across answer crosses every Down answer. Even in this case, the distribution of the final solution size for random thresholds $\{\varphi_x\}$ is not immediately obvious. This section outlines a solution algorithm whose corresponding stochastic process formulation allows insight into the distribution of the final solution proportion.

It is easily seen that once the difficulty thresholds $\{\varphi_x\}$ are determined, the (maximal) final solution set of the puzzle does not depend on the algorithm used to solve it, as long the algorithm fulfills the obvious criterion that it not stop until no more answers may be found. Algorithms may vary in terms of the number of steps they require to reach the solution, but we will not be concerned with that here—all we require is the final solution set. For a puzzle whose grid is a fully white $m \times m$ square, and thus whose crossword network is the complete bipartite network $K_{m,m}$, and whose thresholds $\{\varphi_x\}$ are arbitrary for now, consider the following natural solution algorithm, which alternates Across and Down scans of the puzzle.

1. Determine X_1, the number of Across answers x with $\varphi_x \leq 0$ (i.e., the number of Across seeds).
2. Determine Y_1, the number of Down answers y that can now be found (i.e., that have $\varphi_y \leq X_1/m$).
3. If $Y_1 = 0$, stop—no more answers can be found. Otherwise, if $Y_1 > 0$, then find X_2, the number of *newly* solvable Across answers (i.e., whose threshold φ_x satisfies $0 < \varphi_x \leq Y_1/m$).
4. If $X_2 = 0$, stop. Otherwise, find Y_2, the number of newly solvable Down answers (i.e., with threshold φ_y satisfying $X_1/m < \varphi_y \leq (X_1 + X_2)/m$).
5. Continue in this fashion, letting X_i (resp. Y_i) be the number of Across (resp. Down) answers found at the ith pass in that direction. Once

one of the X_i or Y_i is 0, no more answers can be found. If at this point $\sum X_i = m$ or $\sum Y_i = m$, then the puzzle has been completely solved.

If the thresholds $\{\varphi_j\}_j$ are iid random variables with common cdf $F(t) := P(\varphi_j \leq t)$, then we can set up the following process to give an exact description of the transitions. This process will not be strictly Markov, since it is not "memoryless." However, it does have a restricted memory—see recursions (1) below. Define

$$S_i := \sum_{j=1}^{i} X_j \quad \text{and} \quad T_i := \sum_{j=1}^{i} Y_j$$

to be the sequence of partial sums (i.e., the total number of answers of a given orientation solved by the end of the ith pass). For future convenience, define $T_{-1} := -\infty$, $S_0 = T_0 := 0$. Also denote the final number of solved Across and Down answers by $S_\infty := \sum_{i \geq 1} X_i$ and $T_\infty := \sum_{i \geq 1} Y_i$, respectively; if S_∞ or T_∞ equals m, then the puzzle has been completely solved.

We write $X \sim \text{Bin}(n, p)$ to indicate that X is a binomially distributed random variable with parameters n (number of trials) and p (probability of success on each trial).

For $i \geq 1$, the following distributional recursions hold:

$$X_1 \sim \text{Bin}\big(m, F(0)\big),$$

$$Y_1 | \{X_1\} \sim \text{Bin}\left(m, F(X_1/m)\right), \tag{1}$$

$$X_i | \{S_{i-1}, T_{i-1}, T_{i-2}\} \sim \text{Bin}\left(m - S_{i-1}, \frac{F(T_{i-1}/m) - F(T_{i-2}/m)}{1 - F(T_{i-2}/m)}\right),$$

$$Y_i | \{T_{i-1}, S_i, S_{i-1}\} \sim \text{Bin}\left(m - T_{i-1}, \frac{F(S_i/m) - F(S_{i-1}/m)}{1 - F(S_{i-1}/m)}\right).$$

These recursions can be explained as follows. At, for example, Across-step i, there are $m - S_{i-1}$ remaining Across answers that are not yet solved. The probability of a given answer being newly solvable is the probability that T_{i-2} crossing Down answers were not sufficient to find it, but that T_{i-1} of them were. This is thus the probability that its threshold is less than T_{i-1}/m, *conditioned on* it not being less than T_{i-2}/m. (The values of T_{-1}, S_0, and T_0 are just chosen to make the recursions applicable for $i = 1$ and $i = 2$, and of course we define $F(-\infty) := 0$.) For the scaled processes $\tilde{X}_i = X_i/m$,

$\tilde{Y}_i = Y_i/m$, $\tilde{S}_i = S_i/m$, and $\tilde{T}_i = T_i/m$, the recursions are somewhat neater:

$$\tilde{Y}_1|\{\tilde{X}_1\} \sim \frac{1}{m}\text{Bin}\left(m, F(\tilde{X}_1)\right), \tag{2}$$

$$\tilde{X}_i|\{\tilde{S}_{i-1}, \tilde{T}_{i-1}, \tilde{T}_{i-2}\} \sim \frac{1}{m}\text{Bin}\left(m(1-\tilde{S}_{i-1}), \frac{F(\tilde{T}_{i-1}) - F(\tilde{T}_{i-2})}{1 - F(\tilde{T}_{i-2})}\right), \tag{3}$$

$$\tilde{Y}_i|\{\tilde{T}_{i-1}, \tilde{S}_i, \tilde{S}_{i-1}\} \sim \frac{1}{m}\text{Bin}\left(m(1-\tilde{T}_{i-1}), \frac{F(\tilde{S}_i) - F(\tilde{S}_{i-1})}{1 - F(\tilde{S}_{i-1})}\right). \tag{4}$$

As $m \to \infty$, it is reasonable to expect the above stochastic processes to be well approximated by their deterministic versions, where we take

$\tilde{x}_1 = \mathbf{E}\left[\tilde{X}_1\right]$,

$\tilde{y}_1 = \mathbf{E}\left[\tilde{Y}_1|\tilde{X}_1 = \tilde{x}_1\right]$,

$\tilde{x}_i = \mathbf{E}\left[\tilde{X}_i|\tilde{S}_{i-1} = \tilde{s}_{i-1}, \tilde{T}_{i-1} = \tilde{t}_{i-1}, \tilde{T}_{i-2} = \tilde{t}_{i-2}\right]$, $i = 2, 3, \ldots$

$\tilde{y}_i = \mathbf{E}\left[\tilde{Y}_i|\tilde{T}_{i-1} = \tilde{t}_{i-1}, \tilde{S}_{i-1} = \tilde{s}_{i-1}, \tilde{S}_{i-2} = \tilde{s}_{i-2}\right]$, $i = 2, 3, \ldots$.

Evaluating these expressions, we get a simple recursive system:

$$
\begin{aligned}
\tilde{x}_1 &= F(0) \\
\tilde{y}_1 &= F(\tilde{x}_1) & &= F(\tilde{s}_1), \\
\tilde{x}_2 &= F(\tilde{y}_1) - F(0) & &= F(\tilde{t}_1) - F(0), \\
\tilde{y}_2 &= F(\tilde{x}_1 + \tilde{x}_2) - F(\tilde{x}_1) &= F(\tilde{s}_2) - F(\tilde{s}_1), \\
\tilde{x}_3 &= F(\tilde{y}_1 + \tilde{y}_2) - F(\tilde{y}_1) &= F(\tilde{t}_2) - F(\tilde{t}_1)
\end{aligned}
$$

$\cdots \qquad \cdots \qquad \qquad \cdots$

This system telescopes to give the following very simple system of equations: $\tilde{s}_0 = \tilde{t}_0 = 0$, and

$$\tilde{s}_i = F(\tilde{t}_{i-1}), \quad \tilde{t}_i = F(\tilde{s}_i),$$

or

$$\tilde{s}_i = F(F(\tilde{s}_{i-1})), \quad \tilde{t}_i = F(F(\tilde{t}_{i-1})), \quad \tilde{s}_0 = 0, \quad \tilde{t}_0 = F(0). \tag{5}$$

Since F is an increasing function, the sequences $\tilde{s}_i$ and $\tilde{t}_i$ are increasing, and since they're bounded above by 1, they both converge to the smallest fixed point of F, call it s_*:

$$s_* = \min\{s : F(s) = s\} = \min\{s : G(s) = 0\}, \tag{6}$$

where we define $G(x) := F(x) - x$. For random thresholds $\{\varphi_x\}$ whose cdf F satisfies $F(x) < 1$ for all x (or whose density $f_\varphi(x)$ is > 0 for all x), such as the Gaussian ones we will mostly work with, we have $s_* < 1$, and thus, the (asymptotic) proportion of solved answers will be less than 1. This is to be expected, however, since (for instance) there will be a positive fraction of crossing resistant answers that preclude the possibility of a total solution.

Figure 8.4 shows the results of some exact stochastic simulation of the process (2)–(4); there are two cases of interest that deviate from the deterministic prediction of equation (6):

- If there is some u_* less than, and far from, s_* such that $G(u_*)$ achieves a sufficiently small local minimum for $G(x)$. In this case, extinction of the actual process at values near mu_* may be likely, even though the theory predicts the final solution set to be near ms_*. One therefore might expect a bimodal distribution of final solution sizes, clustered near ms_* and mu_* (Figure 8.4, top left). In this case, the stochasticity creates some smaller solution sizes than predicted deterministically. The key here is that as soon as the process stabilizes over one time step, it remains constant forever; this is one manifestation of the non-Markovian nature of the process.
- If s_* is such that $|G(s_*)| \ll 1$, then there is almost a double root of G' at s_*, and there will be a larger root $v_* > s_*$ of G. In this case, $\{s_i\}$ and $\{t_i\}$ might converge to a value near v_* by skipping over the root at s_* due to random effects. In this case, the distribution is again bimodal (Figure 8.4, top right), but with some values that exceed what is predicted deterministically.

Combining these two cases, we can say that we expect a bimodal distribution of final solution sizes if there is a quasi-double-root u_* of G (i.e., a point $u_* \in [0, 1]$ such that $|G(u_*)|$ and $|G'(u_*)|$ are both small).

Figure 8.4 shows the distribution of final solution sizes for 500 independent instances of the chain (2)–(4) run on a 100×100 grid, with thresholds iid $\varphi_x \sim \text{Normal}(\mu, \sigma)$, for indicated values of μ and σ. Note that the bottom row agrees with the theoretical prediction from equation (6), and the top row displays the bimodal behavior in the two near-critical cases discussed above.

These results are similar to ones found in Gleeson and Cahalane [4], in which the authors also produce a fixed-point equation for cascades on tree-like networks to determine the final proportion of activated nodes.

To find the final distribution of solved answers in the $K_{m,m}$, one should therefore simulate the stochastic process with transitions given by the chain (2)–(4) if m is small, or simply numerically solve the fixed-point equation (6) if m is large, taking care if there is a quasi-double-root of $G(x)$.

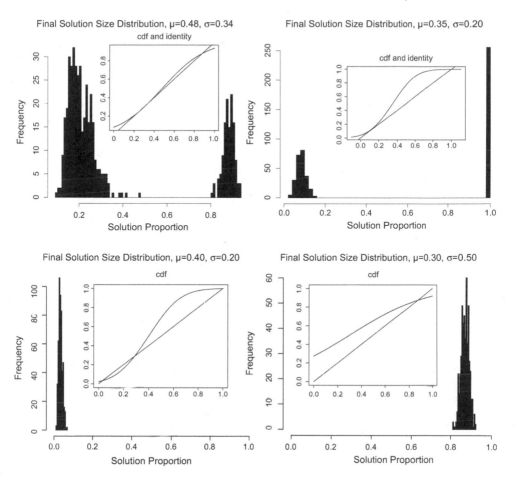

Figure 8.4. Distribution of final solution proportions for 500 simulations on a 100×100 grid with Normal(μ, σ) thresholds, with indicated values of μ and σ. Insets are plots of the corresponding normal cdf F vs. the identity—in both cases on the top row, there is a small quasi-double-root (an actual root in the top left) of $F(x) - x$, and in each case, we see a bimodal distribution. The cases on the bottom are far from critical, and thus the simulations agree with the deterministic predictions.

5 Network Properties and Simulations on Actual Puzzles

This section uses real puzzles from the Sunday *New York Times* [10], builds their crossword networks, and then shows the results of running the cascade solution process on the networks, taking the difficulty thresholds φ_x to be independent Normal(μ, σ) random variables, across a variety of values of μ and σ. We first describe several statistics of the networks that we expect to be relevant to the solution size. We then investigate which features of the output they explain, by comparing simulation results on actual puzzle networks to results on different networks that nevertheless preserve some features of the original newworks.

5.1 Summary Statistics of Actual Crossword Networks

Here we describe a few (low-dimensional) network statistics that are likely relevant to predicting the final solution size.

5.1.1 Clustering. The *clustering coefficient* of a network is a measure of how tightly packed the network is. The usual definition of the clustering coefficient C for a general network is

$$C := 3\frac{n_\triangle}{n_3}, \tag{7}$$

where $n_\triangle$ is the number of triangles, and n_3 the number of ordered connected triples (i.e., paths of length 2) in the network [8]. This quantity C is 1 for any complete network, where every pair of nodes is linked, and any (unordered) triple forms a triangle; C is 0 for any network without loops (i.e., trees) but also for such networks as the two-dimensional lattice. In fact, C is trivially zero for all bipartite networks, and thus for all crossword networks, so we need to amend the definition (7) to obtain something useful for our purposes. Various definitions of the clustering coefficient for bipartite networks have been proposed recently [7, 13], each with its own advantages and level of complexity. For the sake of simplicity, we restrict ourselves to a (scaled) square census of the network: a bipartite graph is highly clustered if it contains as many squares (cycles of length 4) as possible. This quantity has the advantage of being easily computable: if A is the adjacency matrix of the network, and $d_i = \sum_{j=1}^n A_{ij}$ the degree of node i, then the number of squares is

$$n_\square := \frac{1}{8} \sum_{i=1}^n \left((A^4)_{ii} - d_i^2 - \sum_{j \neq i}(A^2)_{ij} \right),$$

which can be explained as follows. The term $(A^4)_{ii}$ counts the number of paths of length 4 beginning and ending at i. Not all of these are true squares, however: we subtract the number of paths of the form $i \to j \to i \to k \to i$, of which there are d_i^2 (j and k can be any neighbors of i, and can be equal to each other); we also subtract paths of the form $i \to j \to k \to j \to i$, and the number of these is the number of paths of length 2 which start at i and end elsewhere. Finally, the factor of $1/8$ is to avoid counting the same square starting at different nodes or in a different direction. The clustering coefficient γ_4 used here is

$$\gamma_4 := \frac{n_\square}{(1/2)\binom{n_a}{2}\binom{n_d}{2}} = \frac{1}{n_a(n_a - 1)n_d(n_d - 1)} \sum_{i=1}^n \left((A^4)_{ii} - d_i^2 - \sum_{j \neq i}(A^2)_{ij} \right), \tag{8}$$

where n_a and n_d are the number of Across and Down answers, respectively. The factor $(1/2)\binom{n_a}{2}\binom{n_d}{2}$ counts the total possible number of squares, and thus $\gamma_4 \in [0, 1]$.

5.1.2 Modularity. Recall from Section 2 that a network has high modularity if its nodes can be partitioned into subsets with many edges within each subset, but few between. There are many ways to assign a scalar value to the modularity of a network, and we will not go into the details of such issues here. We simply use the function `fastgreedy.community` from R's `igraph` package, which is based on the greedy community detection algorithm of Clauset et al. [2]; this function returns a modularity value between 0 and 1. The reader is encouraged to see Clauset et al. [2] for the details; for our purposes, it suffices to note that the larger the modularity score, the more modular the network.

5.1.3 Degree distribution. We record the mean and standard deviation of the degree distribution of each network (i.e., the distribution of answer lengths). These distributions are often bimodal, with one or two large values representing the length of theme answers, and the rest clustered around 4.

5.2 Simulations on Modified Networks

To be able to determine which network statistics are the most relevant, I ran the cascade process on actual crossword networks and also on modifications of these that remove some properties of the original networks. The networks analyzed are:

- Two networks G_1 and G_2 from actual Sunday *New York Times* crossword puzzles [10]. The puzzle grids for each of these were 21×21 squares.
- Rewired networks G_1' and G_2' obtained by keeping the same degree sequence as the G_i, but randomly rewiring the edges in the manner of the so-called "configuration model" network of Molloy and Reed [9]. A way to picture this process is the following. We cut each edge in its middle, leaving each node with a certain number of stubs. We then pair up the stubs randomly to form a new network. In this way, the degree of each node is preserved, but the network is not necessarily bipartite anymore and presumably will lose much of its clustering and modularity.
- Erdős-Rényi random network E with the same average degree as G_1. We can think of E as being made as follows: remove all the edges from G_1 and reattach each end to a randomly chosen new pair of vertices. This has a similar effect to the rewiring, while also slightly reducing degree variability, it also introduces isolated nodes, whose answers can never be found unless their threshold is ≤ 0.
- The network C from the cryptic crossword in Figure 8.3.

TABLE 8.1.
Network statistics for sample crossword puzzles

Parameter	G_1	G_1'	G_2	G_2'	E	C Cryptic
Number of Across and Down answers	64, 74	—	67, 73	—	—	16, 14
Total number of answers n	138	138	140	140	138	30
$\gamma_4 \times 10^3$	0.195	—	0.165	—	—	0.235
Degree mean	5.246	5.246	5.300	5.300	5.246	7.4
Degree standard deviation	2.733	2.733	2.752	2.752	2.616	2.222
Modularity score	0.702	0.379	0.618	0.407	0.386	0.582
μ_{max}	0.25	0.32	0.28	0.33	0.08	0.13

Note: Shown are network statistics for crossword networks G_1, G_2 generated from Sunday New York Times puzzles; rewired configuration-model versions of these, G_1' and G_2'; Erdős-Rényi random network E with same mean degree; and cryptic crossword network C from the grid on the left of Figure 8.3b. Network statistics are explained in Section 5.1; γ_4 is defined in equation (8), and the modularity score is discussed in Section 5.1.2. Since the graphs G_1', G_2' and E are not bipartite, we can define neither the "Across" and "Down" nodes nor the bipartite clustering coefficient γ_4, which is why those entries have no values.

Table 8.1 presents network statistics from Section 5.1 for these networks. In the last line, μ_{max} denotes the largest average difficulty threshold that still allows a full solution to the puzzle (for some value of σ) over several difficulty threshold assignments (see Section 5.3 for more details). In keeping with the idea that an ideal puzzle should be as difficult as possible while still being fully solvable, this μ_{max} is therefore the ideal average clue difficulty level that a puzzle constructor should strive for.

5.3 Simulation Results

Using the `igraph` package in R, I have generated the cascade solution process on the networks G_1, G_1', E, and C mentioned in Section 6.2. For each puzzle, I chose a Normal(μ, σ) distribution for the difficulty thresholds, and over a range of values of both μ and σ, I determined the proportion of answers that were found for each pair (μ, σ). The graphs in Figures 8.5–8.7 in Appendix 1 are color-coded as follows.

- **Red:** 100% of answers found.
- **Orange:** Between 90% and 99% of answers found. In these cases, the puzzle is fully solved, except for one small cluster or a few isolated answers with exceptionally high thresholds.
- **Yellow:** Between 50% and 90% of answers found.
- **Green:** Between 5% and 50% of answers found.
- **Blue:** Less than 5% of answers found. In this regime, there may be some rare answers with low difficulty thresholds that can be found,

but the average threshold is too high to allow any sort of cascade of solutions.

These regions are exhibited in Figures 8.5–8.7 and in Appendix I. The vertical axis is the threshold mean μ, and the horizontal axis its standard deviation σ; both μ and σ are in the range $[0, 0.5]$, at intervals of 0.005 (i.e., 100×100 data points).

5.4 Solution Examples

Figures 8.8–8.12 in Appendix II show a puzzle grid with a few different threshold distributions; those parts of the grid with answers found are filled in green. Habitual crossword solvers may recognize some of these partially solved patterns from stubborn puzzles that they were not able to finish!

5.5 Interpretation of Results

From comparing the first two plots in Figure 8.5, we can see that the reduction in modularity in the rewired network increases the region where a full solution is possible, by removing the possibility of having tightly connected clusters that the solution process fails to reach. This effect can be seen explicitly in the grids in Figure 8.8. Indeed, for a connected network with low modularity, one would expect that the main reason for a node remaining inactive is having a high threshold, rather than being disconnected from a cluster of active nodes.

For the Erdős-Rényi network in Figure 8.6b, a full solution is rare simply because the model allows isolated nodes, which can only be activated if they are seeds from the outset. The profile for the nearly-full solution proportion (orange) is, however, similar to that of the rewired network, indicating that the variation in degree distribution of the original network does not seem to have played a significant role (beyond allowing nodes of degree 0).

Finally, the plot for the cryptic puzzle in Figure 8.7b shows that a cryptic puzzle can be solved only when the average clue difficulty is quite low, at most around 10%. This was, of course, expected, since in cryptic puzzles, the number of letters in an answer is generally strictly greater than the number of crossing answers.

6 Conclusion

Any crossword puzzle enthusiast has no doubt been in the situation of being able to easily solve a puzzle one day, yet be unable to make much headway at all the next day (or week) on a puzzle that is ostensibly of the same level of difficulty. Has the solver suddenly lost the knack for crossword puzzles, or are such occurrences to be expected, and can they be explained quantitatively?

To explain such phenomena, we sought to quantify the dynamics of the solution of a crossword puzzle—in particular, to determine how the puzzle structure and the clue difficulty interact to dictate the overall puzzle difficulty.

This model crucially has a stochastic (random) element, to account for the inevitable variability between how two different solvers might perceive a clue's difficulty; this randomness in turn is responsible for the unpredictable behavior that can be observed in the solutions to these puzzles. The crossword puzzle grid structure is represented by a network or graph object, where crossing answers represented linked nodes, and used a random cascade model inspired by analysis of epidemics or trend diffusions to describe the spread of the solution through the puzzle.

We first considered the case of an artificial puzzle with no black squares, and thus a simple underlying complete bipartite network. In this case, it was shown that the proportion of answers that can be found is well estimated by the solution to a simple fixed-point equation involving the distribution function of the clue difficulty thresholds, unless this fixed point is a near-critical case, in which case the randomness of the clue difficulty has a pronounced effect in the final solution size distribution.

Real crossword puzzles are complex structures that defy such a tidy analytical solution. Instead, the solution process was simulated over actual puzzles and we determined that a real puzzle may be fully solvable, even if the average answer needs about 25% of its letters to be present in order to be found. By comparing results on actual puzzles with results on modified versions thereof, we found that the modularity of the original puzzle is a crucial impediment to achieving a full solution; this is borne out by the simulated solutions in Figure 8.9, for example, where modules of tightly connected answers resist the solution that propagates through the rest of the puzzle.

By conducting such simulations, we are able to determine relationships between the clue difficulty distribution and the final proportion of answers found. From the point of view of a puzzle constructor, this serves to indicate how hard to make the clues to achieve a desired overall puzzle difficulty for some idealized solver; conversely, from the point of view of the solver, this indicates the range of puzzle difficulties for which he or she may expect to be able to solve the puzzle.

Acknowledgments

I thank the organizers and participants of the 2010-11 Program on Complex Networks at the Statistical and Applied Mathematical Sciences Institute (SAMSI; Research Triangle Park, North Carolina) for providing the inspiration for this project. I also thank the anonymous referee for many helpful comments that greatly improved the clarity and flow of this chapter.

Appendix I. Simulations over (μ, σ) Parameter Space

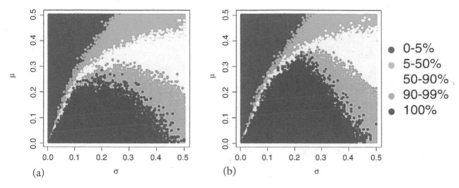

Figure 8.5. Average solution proportion for ten simulations of the cascade process with Normal(μ, σ) thresholds, for each pair of values of (μ, σ) in the range $0 \leq \mu \leq 0.5, 0 \leq \sigma \leq 0.5$, at intervals of 0.005. (a) Standard 21 × 21 Sunday *New York Times* puzzle G_1. (b) Rewired configuration-model network G_1' as described in Section 5.2. A full solution is possible for a wider range of parameters for the rewired network, since much of the modularity of the original network has been removed.

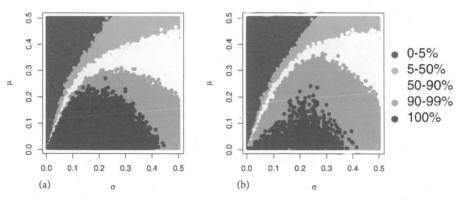

Figure 8.6. Average solution proportion for ten simulations of the cascade process with Normal(μ, σ) thresholds, for each pair of values of (μ, σ) in the range $0 \leq \mu \leq 0.5, 0 < \sigma \leq 0.5$, at intervals of 0.005. (a) standard 21 × 21 Sunday *New York Times* puzzle. (b) Erdős-Rényi random network E with the same number of edges (as described in Section 5.2). Isolated nodes in an Erdős-Rényi network may prevent full solutions even in the case where the parameters make the puzzle easy. However, the orange region is larger for the Erdős-Rényi network, since the modularity has been reduced, as in the rewired network in Figure 8.5.

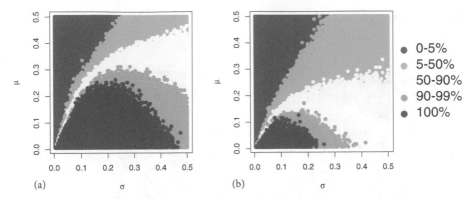

(a) σ   (b) σ

- 0-5%
- 5-50%
- 50-90%
- 90-99%
- 100%

Figure 8.7. Average solution proportion for ten simulations of the cascade process with Normal (μ, σ) thresholds, for each pair of values of (μ, σ) in the range $0 \leq \mu \leq 0.5$, $0 \leq \sigma \leq 0.5$, at intervals of 0.005. (a) standard 21×21 Sunday *New York Times* puzzle; (b) cryptic crossword puzzle C from Figure 8.3. Since the cryptic puzzle is sparser, we see a much smaller set of parameters where a full solution can be found, as expected.

Appendix II. Visualizations of Partially Solved Grids

As noted in Section 5.4, the figures below represent final solutions of a single Sunday *New York Times* puzzle over a variety of parameters for the difficulty threshold, with four instances of the solution for each parameter set. The black squares are the original black squares in the puzzle, the green squares are letters in answers that are solved, and the white squares are letters that cannot be found.

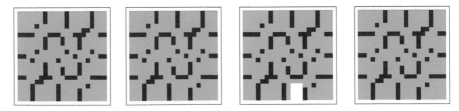

Figure 8.8. Four solution instances on a standard puzzle with Normal(0.2, 0.23) thresholds. These parameter values lie comfortably in the solvable regime, with the third puzzle providing a rare example of a stubborn cluster of answers (white region) stymieing a complete solution.

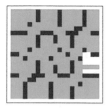 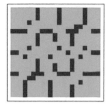 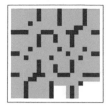 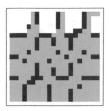

Figure 8.9. Four solution instances on a standard puzzle with Normal(0.25, 0.23) thresholds. The puzzles are mostly solvable, but these cases illustrate well the effect of the modularity of the network: there are pockets of answers (with presumably high thresholds) where the solution does not propagate (white), since these pockets are weakly connected to the rest of the puzzle.

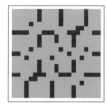  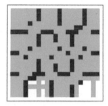 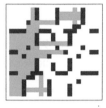

Figure 8.10. Four solution instances on a standard puzzle with Normal(0.3, 0.23) thresholds. Notice in this case the wide variability in the number of answers found, which indicates that these parameters are near a critical threshold.

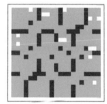 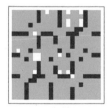 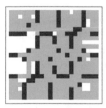 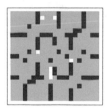

Figure 8.11. Four solution instances on a standard puzzle with Normal(0.25, 0.5) thresholds. As in Figure 8.9, these puzzles are mostly solved, but the obstacles to a full solution are different. In Figure 8.9, the modularity of the network was the main issue. Here, notice the many isolated white squares: because the standard deviation of the thresholds is large at $\sigma = 0.5$, a noticeable number of answers have thresholds so high that they can't be found even with only one letter missing.

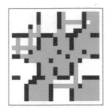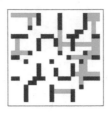

Figure 8.12. Four solution instances on a standard puzzle with Normal(0.4, 0.23) thresholds. By and large, the clues are too difficult in this case, but notice the large variability in the number of answers found. In the second instance, enough of the easy seed answers were found initially to provide a cascade of answers that covered more than half the grid. However, in the fourth instance, too few seed answers intersected in too few places, and the solution did not cascade at all. (This is the situation one would hope for in an epidemiological context!)

References

[1] J. Balogh and B. G. Pittel. Bootstrap percolation on the random regular graph. *Rand. Struct. Alg.* (December 2006) 257–286.

[2] A. Clauset, M. E. J. Newman and C. Moore. Finding community structure in very large networks. *Phys. Rev. E* **70** (2014) 066111.

[3] S. Fortunato. Community detection in graphs. *Phy. Rep.* **486** no. 3–5 (2010) 75–174. DOI: 10.1016/j.physrep.2009.11.002.

[4] J. P. Gleeson and D. Cahalane. An analytical approach to cascades on random networks, in J. Kerstész, S. Bornholdt, and R. N. Mantegna, editors, *Proc. SPIE Noise and Stochastics in Complex Systems and Finance*, Florence, Italy, 66010W, 2007.

[5] M. J. Keeling and K. T. D. Eames. Networks and epidemic models. *J.R. Soc. Interface* **2** (2005) DOI: 10.1098/rsif.2005.0051.

[6] M. J. Keeling. The effects of local spatial structure on epidemiological invasions. *Proc. R. Soc. Lond. B* **266** (1999) 859–867.

[7] P. G. Lind, M. C. González and H. J. Herrmann. Cycles and clustering in bipartite networks. *Phys. Rev. E* **72** (2005) 056127.

[8] M. E. J. Newman. Random Graphs with Clustering. *Physical Review Letters* **103** (2009) 058701.

[9] M. Molloy and B. Reed. A critical point for random graphs with a given degree sequence. *Rand. Struct. Algor.* **6** (1995) 161–180.

[10] W. Shortz, editor. *Everyday Sunday Crossword Puzzles.* St. Martin's Griffin, New York, (2006).

[11] D. J. Watts. A simple model of Global cascades on random networks. *PNAS* **99** no. 9 (2002) 5766–5771.

[12] D. J. Watts and S. H. Strogatz. Collective dynamics of "small-world" networks. *Nature* **393** no. 6684 (1998) 409–10. DOI: 10.1038/30918.

[13] P. Zhang, J. Wang, X. Li, Z. Di and Y. Fan. The clustering coefficient and community structure of bipartite networks. *Physica A* **387** (2008) 27.

9

FROM THE OUTSIDE IN

Solving Generalizations of the
Slothouber-Graatsma-Conway Puzzle

Derek Smith

The Slothouber-Graatsma-Conway puzzle asks you to assemble six $1 \times 2 \times 2$ pieces and three $1 \times 1 \times 1$ pieces into the shape of a $3 \times 3 \times 3$ cube, as shown in Figure 9.1. The puzzle can be quite difficult to solve on a first pass; it has only one solution up to the symmetries of the cube. By contrast, the well-known $3 \times 3 \times 3$ Soma Cube has 240 different solutions.

If you have never tried to solve this puzzle, immediately stop reading and give it a shot! Your local hardware or hobby shop can provide the wooden cubes and glue, or you might find that the puzzle is not too hard to render on paper, or you might be able to find the puzzle at your favorite puzzle shop or online megastore.

The Slothouber-Graatsma-Conway puzzle has been generalized to larger cubes, and it is an infinite family of such puzzles, communicated by Andy Liu, that is the focus of this chapter. Our main result is the following.

Theorem 1. *For any odd positive integer* $n = 2k + 1$, *there is exactly one way, up to symmetry, to make an* $n \times n \times n$ *cube out of* n *tiny* $1 \times 1 \times 1$ *cubes and six of each of the following rectangular blocks:*

$$
\begin{aligned}
1 \times \quad 2k \quad \times 2, \\
2 \times (2k - 1) \times 2, \\
3 \times (2k - 2) \times 2, \\
\vdots \\
k \times \ (k + 1) \ \times 2.
\end{aligned}
$$

Following Perković [8], let us call the $3 \times 3 \times 3$ puzzle (when $k = 1$) the Slothouber-Graatsma-Conway puzzle, combining references to the puzzle's sources in the literature. Certainly, the figures that appear in *Cubics: A Cubic Constructions Compendium* by the architects Graatsma and Slothouber [7],

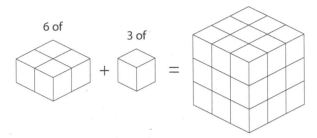

6 of

3 of

Figure 9.1. Can you solve this puzzle?

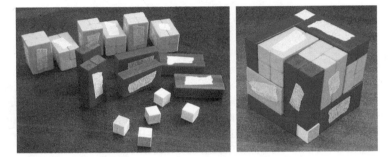

Figure 9.2. The solution to the $5 \times 5 \times 5$ cube.

pp. 82–83, 108–109 show how to assemble the cube, even if these and other constructions in their book are not specifically offered as puzzles. Berlekamp, Conway, and Guy [3], pp. 736–737 describe the puzzle, unnamed but attributed to one of the book's coauthors, with the $1 \times 1 \times 1$ pieces considered as empty holes left by the other six pieces. The puzzle is given different names—such as "Slothouber-Graatsma puzzle" [4], "Conway's Curious Cube" [5], and "Pack It In" TM—in different books and online puzzle resources.

The $5 \times 5 \times 5$ puzzle in our family (when $k = 2$) is not to be confused with another $5 \times 5 \times 5$ packing puzzle due to Conway, variously called "Blocks-in-a-Box" [3], "Conway's puzzle" [4], and "Conway's Cursed Cube" [5]. That puzzle, which Conway created along with our $3 \times 3 \times 3$ puzzle when he was a student at Cambridge, consists of three $1 \times 1 \times 3$ rods, one $1 \times 2 \times 2$ square, one $2 \times 2 \times 2$ cube, and thirteen $1 \times 2 \times 4$ planks. Our $5 \times 5 \times 5$ puzzle is also attributed to Conway [9], but Conway shares credit for the puzzle with O'Beirne [6]. This puzzle has been produced and sold under the names "Made to Measure" and "Shipper's Dilemma" (it is one of several puzzles to be given the latter name). It was offered as Problem 213 in the problem section of *Math Horizons* in February 2008, communicated by Ben Bielefeld [2]. Following a hint given in the September issue [1], Mark Sand's Special Topics class at Dana College (Blair, Nebraska) found the unique solution, as evidenced by the model the class created from colorful Cuisenaire® rods taped together (Figure 9.2).

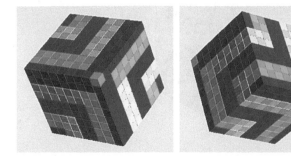

Figure 9.3. Two views of the solution to the $9 \times 9 \times 9$ cube.

The $7 \times 7 \times 7$ puzzle in our family is attributed to Conway and Thiessen in Jerry Slocum mechanical puzzle collection (http://webapp1.dlib.indiana.edu/images/splash.htm?scope=lilly/slocum)

The next section describes a way to solve each puzzle in the family and explains why there are no other solutions. The final section presents several related open problems.

1 Placing Pieces

The two images in Figure 9.3 are taken from BurrTools, a fabulous free software package by Andreas Röver that solves packing puzzles of various types. The images show the solution of the $9 \times 9 \times 9$ puzzle from two different perspectives, exhibiting the three-fold rotational and antipodal symmetries that are present in all solutions.

The main theorem is proved by means of a sequence of lemmas that restrict the placement of certain pieces, starting with the smallest and thinnest. We begin the proof by restricting the placement of the $1 \times 1 \times 1$ *tiny cubes*. We define a *slice* of an $n \times n \times n$ cube as any $n \times n$ square of cells a fixed distance from one of the six outer faces of the cube. For example, Figure 9.4 shows an orange slice of a $5 \times 5 \times 5$ cube at distance 2 from the back left face.

Lemma 1 (Slice). *Each slice contains exactly one tiny cube.*

Proof. Each of the slices of the cube is a square of size $n \times n$. Since n is odd, a slice contains an odd number of cells. Notice that apart from the tiny cubes, each puzzle piece has two even side lengths, so it must intersect each slice in an even number of cells, regardless of its orientation with respect to the slice. For example, the blue $2 \times 2 \times 3$ piece in Figure 9.4 intersects the orange slice in a 2×2 square. Thus, each slice must contain an odd number of tiny cubes, and so it must contain at least one of them. Since there are only n tiny cubes to go around, there must be exactly one in each slice. □

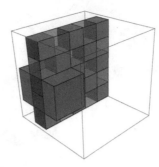

Figure 9.4. A $2 \times 3 \times 2$ blue piece intersecting a checkerboard-colored orange slice in a $5 \times 5 \times 5$ cube.

While it is not needed for the rest of the proof, note that any piece except a tiny cube must intersect a slice in the same number of shaded and unshaded cells of a checkerboard coloring of the slice. For example, the blue piece in Figure 9.4 intersects two shaded and two unshaded cells of the slice. This further restricts the placement of each tiny cube, because in a checkerboard coloring of the entire cube, it must occupy a cell that is the same color as the corner cells of each slice that contains it.

The next step is to restrict the flat $1 \times 2k \times 2$ pieces to the outer faces of the cube.

Lemma 2 (Flat). *Each face of the cube has exactly one $1 \times 2k \times 2$ piece lying flat on it.*

Proof. By Lemma Slice, there exists a tiny cube in the second slice away from each of the six outside faces. The single cell between this tiny cube and the outside face can only be filled by a $1 \times 2k \times 2$ piece: it can't be filled by another tiny cube (doing so would put two tiny cubes in the same slice), and all of the other pieces are too thick. □

We can now, finally, place a couple of tiny cubes in the outermost slices of the cube.

Lemma 3 (Corners). *The only tiny cubes in the outside slices of the cube are two tiny cubes in opposite corners of the cube.*

Proof. No tiny cube can be in the interior of an outside slice or in the middle of one of its edges. To see this, consider Figure 9.5a, a bird's-eye view of a tiny cube attempting to occupy the yellow cell on the bottom slice of a cube of arbitrary size. What pieces can occupy the adjacent cells labeled a, b, c, and d? By Lemma Slice, no other tiny cube can do it; and by Lemma Flat, at most one

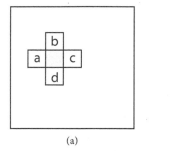
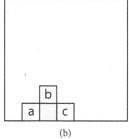

(a)   (b)

Figure 9.5. In this bird's-eye view, the cell directly above a tiny cube in yellow cannot be filled. (a) Bottom slice; (b) edge of bottom slice.

of the four cells can be occupied by a $1 \times 2k \times 2$ piece lying flat on the bottom face. Thus, the other three labeled cells must be occupied by pieces of height at least 2 above the bottom face. But this would leave an unfillable cell directly above the tiny cube in yellow.

The same type of reasoning applies in Figure 9.5b, with the tiny cube attempting to occupy the yellow cell on the edge of a slice. In this case, one of cells a and c could be part of a $1 \times 2k \times 2$ piece lying flat against the southern face, but as before this would still lead to an unfillable gap above the tiny cube in yellow.   □

With two corner cells now occupied by tiny cubes, we can place all of the $1 \times 2k \times 2$ pieces.

Lemma 4 (Propeller). *Three $1 \times 2k \times 2$ pieces surround each of the two corner tiny cubes, as shown in Figure 9.6b.*

Proof. There are three cells adjacent to each corner tiny cube. If a piece thicker than a $1 \times 2k \times 2$ piece occupies one of these three cells, at most one of the other two cells could be occupied (necessarily by a $1 \times 2k \times 2$ piece). So these three adjacent cells must be occupied by three $1 \times 2k \times 2$ pieces lying flat against three different outer faces of the cube.

For $k > 1$, there are only two possible ways to place each of these three pieces up to symmetry, and Figure 9.6a shows why one of these can't work: the red cell between the piece and right front face can't be filled, because none of the three $1 \times 2k \times 2$ pieces adjacent to the tiny cube in the opposite corner can reach it. The propeller shape displayed in Figure 9.6b is then the only possible configuration, up to reflecting the propeller's orientation.   □

When $k = 1$, the $1 \times 2k \times 2$ pieces are $1 \times 2 \times 2$ and so have more symmetry than when $k > 1$. Then there is really only one way, not two, to place such a

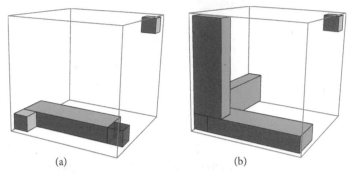

Figure 9.6. There is only one correct way to place a $1 \times 2k \times 2$ piece next to a corner tiny cube. (a) Does not work; (b) correct placement.

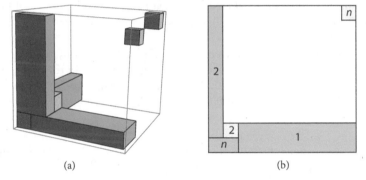

Figure 9.7. Two views of the pieces that have been forced prior to Lemma 5. (a) Three-dimensional view; (b) bird's-eye view.

piece next to a tiny corner cube. Placing all six pieces this way, and putting the final tiny cube in the central cell, completes the $3 \times 3 \times 3$ puzzle.

For $k > 2$, we cannot yet place the propeller of three $1 \times 2k \times 2$ pieces adjacent to the other corner tiny cube, because we have not yet ruled out either of its two possible orientations. Regardless of which orientation is correct, we are forced to place two more tiny cubes as shown in Figure 9.7a, because no other remaining piece can fill either of those two cells without leaving an unfillable gap of width 1 next to one of the three nearby outer faces. When $k = 2$, the second propeller must be oriented antipodally with respect to the first one to allow the six $2 \times 3 \times 2$ pieces to fill in the cells between the two propellers (in only one way), completing the $5 \times 5 \times 5$ puzzle.

The next few diagrams are drawn as bird's-eye views to more easily display the possible placements of important pieces. The labels on the pieces are heights above the bottom face, with all the cells below a labeled piece occupied.

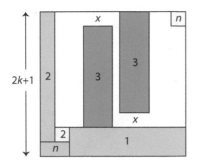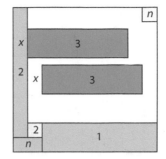

Figure 9.8. The $3 \times (2k - 2) \times 2$ pieces can't go in these positions.

Figure 9.7b has an n in the upper right corner, because, regardless of which orientation is chosen for the propeller around that tiny cube, the column under that cell is now occupied.

In the final lemma, we place the $3 \times (2k - 2) \times 2$ pieces and then the $2 \times (2k - 1) \times 2$ pieces. I then explain why this lemma is essentially the final one we need.

Lemma 5 (Three). *For $k > 2$, each outside face of the cube has a $3 \times (2k-2) \times 2$ piece on it. The piece is oriented to have height 3 away from the face, and it is adjacent to a $1 \times 2k \times 2$ piece already placed. This forces the placement of the $2 \times (2k - 1) \times 2$ pieces as well.*

Proof. By Lemma Slice, there exists a tiny cube in the fourth slice away from each outside face. Due to the four tiny cubes already placed, this tiny cube must lie in the inner $(n - 4) \times (n - 4) \times (n - 4)$ cube. The three cells between this tiny cube and the outer face can be filled only with a $3 \times (2k - 2) \times 2$ piece: all other remaining piece dimensions are either larger than 3 or are even.

Where does the $3 \times (2k-2) \times 2$ piece lie on the bottom face? Not in the four types of placements shown in Figure 9.8: those leave unfillable gaps of width 1 in the spaces marked with an x.

So the only possible placement of the $3 \times (2k - 2) \times 2$ piece is against the right wall, as shown in Figure 9.9a. Note the gap of width 3 above height 2 on the far left: this gap can only be filled with another $3 \times (2k-2) \times 2$ piece, which must be placed in the position outlined with dashes to avoid unfillable gaps of width 1.

The outlined $3 \times (2k-2) \times 1$ piece rests on top of a $1 \times 2k \times 2$ piece. By the symmetry of the pieces already placed, this forces all $3 \times (2k - 2) \times 2$ pieces to be adjacent to $1 \times 2k \times 2$ pieces in the same way. Two such pieces are shown in Figure 9.9b.

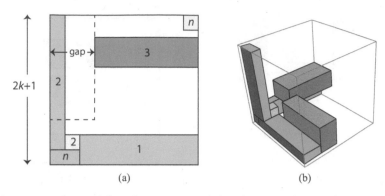

Figure 9.9. The $3 \times (2k - 2) \times 2$ pieces must be adjacent to the $1 \times 2k \times 2$ pieces.

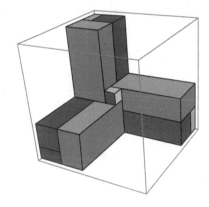

Figure 9.10. The three thinnest types of pieces have now been placed.

With the $3 \times (2k - 2) \times 1$ pieces now placed, the only way to fill in the gaps made by them is with tiny cubes and the red $2 \times (2k - 1) \times 2$ pieces as shown in Figure 9.10. □

When $k = 3$, filling in the central cell of the cube with the remaining tiny cube and choosing the antipodal orientation for the opposite corner is the only way to complete the $7 \times 7 \times 7$ cube. When $k = 4$, the antipodal orientation is also the only orientation that allows the $9 \times 9 \times 9$ cube to be filled in with the six $4 \times (2k - 3) \times 2$ pieces, in exactly one way.

Finally, here's the wonderful punchline to finish the proof of Theorem 1: replicate Lemma "Three" as Lemma "Five", and then as Lemma "Seven", and so on, up to Lemma "k" if k is odd or Lemma "$k - 1$" if k is even. To replicate, simply replace instances of the number 3 throughout the lemma with the number 5, and then with the number 7, and so on. (Of course, you'll also need to alter related numbers by 2 at each step as well.) The point is that configurations of pieces already placed form thicker propellers, allowing an identical argument to be made in each subsequent version of the lemma. For example, in the proof of Lemma "Five", the $5 \times (2k-4) \times 2$ pieces are forced to be adjacent to the $3 \times (2k-2) \times 2$ pieces already placed, and the $4 \times (2k-3) \times 2$ pieces will fill the gaps.

2 Puzzling On

Several open questions immediately come to mind. Are there other interesting infinite families of packing puzzles whose solutions are unique in a nontrivial way? Can some of the pieces we used in our family be subdivided, but still have the puzzle admit only one solution up to symmetry? And is there a snazzier proof of the uniqueness result just presented? Here's a nice fact: if the outer cells of the solution to the $n \times n \times n$ cube are shaved off, what's left is the unique solution to the $(n-2) \times (n-2) \times (n-2)$ cube. But I have so far been unable to use this idea as the basis of an alternate proof, one that works inside out instead of outside in.

There is a further generalization of the Slothouber-Graatsma-Conway puzzle that allows for even side lengths. For $n > 2$, make an $n \times n \times n$ cube with n pieces of size $1 \times 1 \times 1$ and three of each of the following rectangular blocks:

$$1 \times (n-1) \times 2,$$
$$2 \times (n-2) \times 2,$$
$$3 \times (n-3) \times 2,$$
$$\vdots$$
$$(n-1) \times 1 \times 2.$$

Notice that when n is odd, the pieces are exactly the same as those in the family considered earlier. When n is even, you get six of most of the pieces but only three pieces of size $(n/2) \times (n/2) \times 2$. Our uniqueness proof for odd n doesn't even get off of the ground for even n: Lemma Slice is no longer true, because the slices now have an even number of cells, and half of the pieces can intersect slices in an odd number of cells. There is no longer a unique solution for each n. BurrTools shows that there are 43 solutions to the $4 \times 4 \times 4$ puzzle, up to symmetry, and 173 solutions to the $6 \times 6 \times 6$ puzzle. It would be nice to know how the number of solutions grows as n increases.

Acknowledgments

Thanks go to Ben Bielefeld for introducing us to the $5 \times 5 \times 5$ puzzle; to Andy Liu for introducing me to the generalizations for odd n; and to Bill Cutler and his puzzle-solving software for first assuring me several years ago that there is only one solution to the $7 \times 7 \times 7$ puzzle.

References

[1] Anon. Hint in Problem Section. *Math Horizons* **16** no.1 (2008) 33.

[2] B. Bielefeld. Problem 213 in Problem Section. *Math Horizons* **15** no. 3 (2008) 32.

[3] E. R. Berlekamp, J. H. Conway, and R. K. Guy. *Winning Ways for Your Mathematical Plays.* Academic Press, Waltham, MA, 1982. (The second edition of *Winning Ways* was published by A K Peters in 2004.)

[4] S. T. Coffin. *The Puzzling Word of Polyhedral Dissections.* Recreations in Mathematics. Oxford University Press, Oxford, 1990.

[5] S. T. Coffin. *Geometric Puzzle Design.* A K Peters/CRC Press, Boca Raton, FL, 2007.

[6] J. H. Conway. Private communication, 2007.

[7] W. Graatsma and J. Slothouber. *Cubics: A Cubic Constructions Compendium.* Cubics Constructions Centre, Heerlen, the Netherlands, 1970.

[8] M. S. Perković. *Famous Puzzles of Great Mathematicians.* American Mathematical Society, Providence, RI, 2009.

[9] P. van Delft and J. Botermans. *Creative Puzzles of the World.* Key Curriculum Press, Berkeley, CA, 1993.

PART III

◇◇◇◇◇◇◇◇◇◇◇◇◇◇◇◇

Playing Cards

10

◇◇

GALLIA EST OMNIS DIVISA IN PARTES QUATTUOR

Neil Calkin and Colm Mulcahy

Dedicated with great fondness to the memory of visionary publisher Klaus Peters, staunch supporter of recreational mathematics. He was always insightful and practical, yet unfailingly supportive of good ideas and young talent. He also cheerfully provided the title of this paper.

In 2003, the second author discovered a certain move that could be performed on packets of cards which, when repeated four times, brought the cards back to their initial arrangement. That is, the move was "of period four." The move gave rise to some interesting card effects, especially when applied just three times [1]. At that time, the easiest known method for explaining how the move worked involved a decomposition of the packet into three parts.

A decade later, a family of generalizations was stumbled upon, also of period four, with interesting consequences when applied only twice. This was announced and applied in Chapter Nine of the second author's book *Mathematical Card Magic: Fifty-Two New Effects* [2]. However, the only proof given there was for one extreme case, and it was not an extension of any known proof in the older scenario; a different decomposition of the packet into three parts was used.

Here, we present a single unified proof, covering the whole spectrum of cases. It divides packets into four parts.

1 The Ice Cream Trick and What's Behind It

We begin by reviewing the original 2003 move in the magical "Low-Down Triple Dealing" context in which it first appeared [1].

A quarter of a deck of cards is handed to a spectator, who is invited to shuffle freely. She is asked to call out her favourite ice-cream flavor; let's suppose she says, "Chocolate." Take the cards back in one hand, face down, and use the other hand to pull off cards, one at a time (hence, reversing their order) for each letter of "chocolate," to represent a scoop of ice cream, before dropping the remainder of the packet on top (as a "topping").

This spelling/dropping routine—effectively reversing some cards and cutting them to the bottom—is repeated twice more, for a total of three times. Emphasize how random the cards must now be, since they were shuffled before you handled them, and you had no control over the named ice-cream flavor.

Have the spectator press down hard on the card that ends up on top, requesting that she magically turn it into a specific one of your naming, say the Four of Diamonds. When it's turned over it will be found to be the named card.

This is an in-hand version of the first effect in "Low-Down Triple Dealing," the inaugural *Card Colm* published online at MAA.org [1] to mark Martin Gardner's ninetieth birthday. There the spelled-out cards were dealt to a table.

Clearly, you know in advance what card ends up on top. How is this possible? The answer is simple: it's the card that started out at the bottom, and hence you must get a peek at this after you take the cards back from the spectator. The easiest way to do this without arousing suspicion is perhaps to square up the quarter deck, by tapping it on the table, at such an angle that you (and only you) catch a glimpse of the bottom card.

What remains is entirely mathematical: three rounds of this spelling and dropping always brings the original bottom card to the top, subject to an easily satisfied condition. There's a key relationship between the size of the quarter deck and the length of the word spelled out: the effect works as claimed, provided that at least half the cards in the packet are spelled out, and hence reversed.

Hence, if the quarter deck contains twelve to fourteen cards, any flavor of seven to ten letters (such as the ever-popular vanilla, chocolate, or strawberry) works well. However, if only eleven cards are used, it's best to avoid words of ten letters (e.g., strawberry; can the reader see why?), whereas if fifteen cards are used, then vanilla is problematic. An alternative presentation is to ask first for a desired ice cream flavor, and then choose the packet size accordingly. If peach is selected, the packet ideally has from eight to ten cards, but if mint chocolate chip is selected instead, use about half of a regular fifty-two-card deck.

More generally, suppose we start with a packet of n cards, numbered 1 to n from the top down. Fix $k \geq \frac{n}{2}$; for performance reasons, we'll assume that $k \leq n - 2$. The dealing out (hence reversing) of k cards from the packet

$$\{1, 2, \ldots, k - 1, k, k + 1, k + 2, \ldots, n - 1, n\},$$

and dropping the *bottom* $n - k$ cards on top as a unit, yields the rearranged packet

$$\{k + 1, k + 2, \ldots, n - 1, n, k, k - 1, \ldots, 2, 1\}.$$

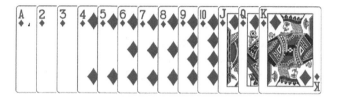

Figure 10.1. A–K♢ fanned face up from left to right.

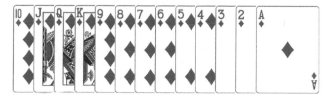

Figure 10.2. A–K♢ after nine cards are reversed and cut.

From this perspective, it's far from clear why repeating this twice more brings the original bottom card n to the top.

Let $n = 13$ and $k = 9$, and suppose we start with the Diamonds in order, face down. Figure 10.1 shows these cards fanned face up, from left to right.

If nine cards are reversed and cut to the bottom, we get what's shown in Figure 10.2. Note that the cards shown on the left in all such displays represent the top of the packet. In performance, the packet would be closed up and face down.

It helps to break up the first k cards 1–k into two parts T (top) and M (middle) of lengths $n - k$ and $k - (n - k) = 2k - n \geq 0$, respectively. (If $n = 2k$, i.e., if exactly half the packet is dealt each time, then M is nonexistent, which is not a problem.) Denoting the final $n - k$ cards by B (bottom), the whole packet thus decomposes symmetrically into the three parts T, M, B, given by

$T = \{1, 2, \ldots, n - k\}$,

$M = \{n - k + 1, n - k + 2, \ldots, k\}$,

$B = \{k + 1, k + 2, \ldots, n\}$.

In the example above, we have $T = \{1, 2, 3, 4\}$, $M = \{5, 6, 7, 8, 9\}$, and $B = \{10, J, Q, K\}$.

In general, it's not hard to see that dealing out k cards and dropping the other $n - k$ on top amounts to the transformation

$$T, M, B \rightarrow B, \overline{M}, \overline{T},$$

where the bar indicates a complete subpacket reversal. A second round of counting and dropping yields $B, \overline{M}, \overline{T} \to \overline{T}, M, \overline{B}$, since two reversals of a subpacket restore it to its initial order. A third round results in $\overline{T}, M, \overline{B} \to \overline{B}, \overline{M}, T$, putting the original bottom card on top, as claimed. On closer inspection, we see that at least half the original packet—the bottom k cards, namely M and B together—ends up on top at the end, but in reverse order.

Obviously, one more round of spelling (i.e., counting) and dropping above will restore the original top card to the bottom. Actually, much more holds: under a fourth such move, we get $\overline{B}, \overline{M}, T \to T, M, B$. The entire packet is now in the same order in which it started out, so that the move in question has period four.

We now formalize this, using slightly different language; the dropping of the remaining cards on top that we spoke of above is replaced by the equivalent cutting of the reversed cards to the bottom.

> **The Bottom to Top and Period-Four Principles.** *If the top k cards of a packet of size n are reversed, where $k \geq \frac{n}{2}$, and those are then cut to the bottom, then every card in the packet is returned to its original position when this move is done three more times. Just doing it three times total brings the original bottom k cards to the top, in reverse order.*

In its original presentation [2], the acronym "COAT" is used to refer to this Counting Out And Transferring cards from top to bottom move, but in what follows, we ignore that; we wish to reserve the letter "c" for cutting.

One very quick way to reverse a portion of cards is just to flip it over as a unit. Interspersed with cutting, this results in a packet of mixed face-up and face-down cards. This is treated in Mulcahy [2, Chapter 9].

It's natural to ask whether the above results generalize in any way, to related moves of period four, and whether there is a corresponding move that brings the original top card (if not more) to the bottom, if done three times. These questions are addressed in the next section.

2 Reversing More, Cutting Less

Earlier we fixed a packet of size n, selected some $k \geq \frac{n}{2}$, and reversed the top k cards in the packet, before effectively cutting all of them to the bottom. Now, let's consider reversing k cards (in situ) and then cutting only $n - k$ of them to the bottom. It's a less smooth move, as it undeniably comes in two distinct phases, and it also implicitly requires full knowledge of the packet size.[1]

---

[1] The move used in the Ice Cream Trick works when at least half the cards are spelled out each time; we don't need to know exactly how many cards form the topping.

Figure 10.3. A–K◇ after nine cards are reversed and four are cut.

Figure 10.4. After nine more cards are reversed and four more are cut.

Figure 10.5. After a third round of reversing nine cards and cutting four.

Reversing the top k cards from the packet $\{1, 2, \ldots, n\}$ in place and then cutting $n - k$ of those from top to bottom yields the packet

$$\{2k - n, 2k - n - 1, \ldots, 2, 1, k + 1, k + 2, \ldots, n - 1,$$
$$n, k, k - 1, \ldots, 2k - n + 2, 2k - n + 1\}.$$

We're ignoring the case $k = n$, since no cards are cut, and also the case $k = \frac{n}{2}$, for which $n - k = k$; those are incarnations of Low-Down Triple Dealing.

Let's look at a specific example. Once more, assume $n = 13$ and $k = 9$, and start with the face-down Diamonds in order. Reversing the top nine cards in place and then cutting four of those to the bottom results in Figure 10.3 when the cards are fanned face up, from left to right.

Repeating this yields Figure 10.4; the original top five cards are back where they started. A third application of the move leads to Figure 10.5.

Unlike in the case of cutting all nine reversed cards to the bottom, we see that it's after two, as opposed to three, rounds of the revised move that we obtain something with magic potential. The good news is that a fourth application once more restores the whole packet to its initial state; something which, with hindsight, can be deduced from the state of the above packet after two rounds of reversing and cutting.

In Mulcahy [2, Chapter 9], this modified move is called "COAT(ML)ing," for Counting Out And Transferring (More and Less cards, respectively). There, "more" and "less" (namely, k and $n - k$) sum to the packet size.

To understand what is going on, we use a different packet decomposition from that employed earlier. Given a packet of n cards 1-n and some fixed $k > \frac{n}{2}$, write

$$n = k+(n-k) = (k-(n-k))+(n-k)+(n-k) = (2k-n)+(n-k)+(n-k),$$

and break the packet into three parts X, Y, Z, of respective sizes $2k - n$, $n - k$, and $n - k$, as follows:

$X = \{1, 2, \ldots, 2k - n - 1, 2k - n\},$

$Y = \{2k - n + 1, 2k - n + 2, \ldots, k - 1, k\},$

$Z = \{k + 1, k + 2, \ldots, n - 1, n\}.$

To quote Julius Caesar, *Gallia est omnis divisa in partes tres*. The smallest X can be is $\{1\}$, which happens if and only if $n + 1 = 2k$ with n odd; when n is even X is at least $\{1, 2\}$. In the example above, we have $X = \{1, 2, 3, 4, 5\}$, $Y = \{6, 7, 8, 9\}$, and $Z = \{10, J, Q, K\}$.

In general, it's not hard to see that reversing the top k cards in place and then cutting $n - k$ of them to the bottom amounts to the transformation

$$X, Y, Z \to \overline{X}, Z, \overline{Y},$$

where the bar, as usual, indicates a complete subpacket reversal. A second round of reversing and cutting leads to $X, \overline{Y}, \overline{Z}$, thus restoring the top $2k - n$ cards to their original positions. Note that up to 100% of the packet may be preserved here, because not only can $2k - n$ be large, but also in the extreme case $k = n - 1$, we see that X contains $n - 2$ cards, whereas Y and Z each contain one card. Hence, the final state $X, \overline{Y}, \overline{Z}$ after two of these moves is actually the same as the initial state X, Y, Z.

A third move takes $X, \overline{Y}, \overline{Z}$ to $\overline{X}, \overline{Z}, Y$, and a fourth takes that back to the starting arrangement X, Y, Z. Hence, we get the following result.

The Top to Top and Period-Four Principles. *If the top k cards of a packet of size n are reversed in place, where $k > \frac{n}{2}$, and then $n - k$ of those are cut to the bottom, then every card in the packet is returned to its original position when this move is done three more times. Just doing it twice in total restores the original top $2k - n$ cards—if not more—to the top, in order.*

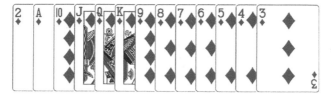

Figure 10.6. A–K♢ after nine cards are reversed and seven are cut.

Figure 10.7. After nine more cards are reversed and seven more are cut.

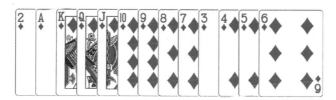

Figure 10.8. After a third round of reversing nine cards and cutting seven.

3 What Have the Romans Ever Done for Us?

So now we know what happens to a packet of size thirteen if we repeatedly reverse the top nine cards in place and then cut either nine or four cards to the bottom. It's natural to ask what happens if we reverse nine each time, but cut some other, intermediate, fixed number of cards, say, seven.

For $n = 13, r = 9$, and $c = 7$, again start with the face-down Diamonds in order, and reverse the top nine cards in place before cutting seven of those to the bottom. Figure 10.6 shows the results, when the cards are fanned face up, from left to right.

Repeating this yields Figure 10.7. This time, we see that the original top two cards are back where they started. A third application of the move leads Figure 10.8.

It should come as no surprise that a fourth application restores the whole packet to its initial state. This strongly suggests the following result, which is stated without proof in Mulcahy [2, Chapter 9].

The General Period-Four Reverse and Cut Principles. *If the top r cards of a packet of size n are reversed in place, where $r \geq \frac{n}{2}$, and c of those are cut to the bottom, where $n - r \leq c \leq r$, then every card in the packet is returned to its original position when this move is done three more times. If $r > \frac{n}{2}$, then doing it just twice in total restores the original top $2r - n$ cards—if not more—to the top, in order. If $r = \frac{n}{2}$, then the original bottom card is on top after three such moves.*

The packet decomposition and proof given above for the case $c = n - r$ doesn't easily adapt to work for any c such that $n - r < c < r$, never mind when $c = r$. Of course, that last special case is in the context of the original Low-Down Triple Dealing, for which a different decomposition and proof has already been given.

We now provide a brand new packet decomposition of 1–n that permits a single unified proof in all cases. The secret is to divide the packet into four parts—it's a case of *Gallia est omnis divisa in partes quattuor:*

$S = \{1, 2, \ldots, r - c\},$
$P = \{r - c + 1, \ldots, n - c\},$
$Q = \{n - c + 1, \ldots, r\},$
$R = \{r + 1, \ldots, n\}.$

Note that S has length $r - c$, and Q has length $(r + c) - n$, whereas P and R both have length $n - r$. Also, S, P, Q together constitute the cards that are reversed in the first round, and P and Q together (reversed) are the ones that are then cut.

Here, $r = c$ precisely when S is nonexistent: that's the Low-Down Triple Dealing situation, with P, Q, R being the first (symmetric) decomposition T, M, B seen in Section 2. Also, $r + c = n$ precisely when Q is nonexistent, with S, P, R then being the second (asymmetric) decomposition X, Y, Z seen in Section 3.

In the example with $n = 13, r = 9$, and $c = 7$, we have

$$S = \{1, 2\}, \qquad P = \{3, 4, 5, 6\}, \qquad Q = \{7, 8, 9\}, \qquad R = \{10, J, Q, K\}.$$

For added clarity, let's break down the reversing of r cards (in place) and subsequent cutting of c of those to the bottom into two stages, as follows.

$$S, P, Q, R \rightarrow \overline{Q}, \overline{P}, \overline{S}, R, \rightarrow \overline{S}, R, \overline{Q}, \overline{P}.$$

Consequently, after a second such composite move, we have

$$\overline{S}, R, \overline{Q}, \overline{P} \rightarrow Q, \overline{R}, S, \overline{P} \rightarrow S, \overline{P}, Q, \overline{R}.$$

And a third such composite move yields

$$S, \overline{P}, Q, \overline{R} \rightarrow \overline{Q}, P, \overline{S}, \overline{R} \rightarrow \overline{S}, \overline{R}, \overline{Q}, P.$$

Finally, after a fourth we get

$$\overline{S}, \overline{R}, \overline{Q}, P \to Q, R, S, P \to S, P, Q, R.$$

For card magic applications of these extended principles, see Mulcahy [2, Chapter 9].

4 Dualism—It's All in How You Group It

Let's return to a question left unanswered earlier: is there a period-four move that brings the original top card (if not more) to the bottom, when done three times? There most certainly is! One approach—without resorting to awkward moves, such as manipulating cards at the bottom of the packet—is to cut cards from top to bottom before reversing in place some of the resulting top cards, instead of the other way around. The number of cards to be cut at the outset needs to be adjusted accordingly, however.

Starting with a packet of size n, and k satisfying $k \geq \frac{n}{2}$, first cut $n - k$ cards from top to bottom, then reverse the new top k cards in place. A little experimentation reveals that doing this three times takes the original top k cards to the bottom, suitably reversed, and a fourth such move gets us back to the initial state.

In a similar way, the other variations of Low-Down Triple Dealing considered earlier adapt to this context. The packet decompositions and proofs provided before in the situations where reversing preceded cutting can all be modified as desired, yielding the following results.

The Top to Bottom and Period-Four Principles. *If the top $n-k$ cards of a packet of size n are cut to the bottom, where $k \geq \frac{n}{2}$, and the resulting k top cards are reversed in place, then every card in the packet is returned to its original position when this move is done three more times. Just doing it three times total brings the original top k cards to the bottom, in reverse order.*

The Second Top to Top and Period-Four Principles. *If the top k cards of a packet of size n are cut to the bottom, where $k > \frac{n}{2}$, and the resulting k top cards are reversed in place, then every card in the packet is returned to its original position when this move is done three more times. Just doing it twice in total restores the original top $2k - n$ cards—if not more—to the top, in order.*

The General Period-Four Cut and Reverse Principles. *If the top c cards of a packet of size n are cut to the bottom, where $r \geq \frac{n}{2}$ and $n - r \leq c \leq r$, following which the resulting top r cards are reversed in place, then every card in the packet is returned to its original position when this move is done three more times.*

If $r > \frac{n}{2}$, then doing it just twice in total restores the original top $2r - n$ cards—if not more—to the top, in order. If $r = \frac{n}{2}$, then the original top r cards are on the bottom, in reverse order, after three such moves total.

There's another way to see why we get period four when we first cut and then reverse, subject to the conditions on n, c, and r above, simply because it's already been shown to hold when we reverse before we cut. Both cutting and reversing are operations that are easily undone, so that they are permutations (i.e., elements of the large algebraic group of all possible rearrangements of n objects). If we denote cutting c cards by C, and reversing r cards (in place) by R, then what we showed earlier was that $C R$ (reversing then cutting) is a permutation of period four, namely, $(C R)(C R)(C R)(C R) = I$ is the identity permutation (which doesn't change anything). Consequently, $C(RC)(RC)(RC)R = I$, and so $C(RC)(RC)(RC)(RC) = C$, from which it follows that $(RC)(RC)(RC)(RC) = I$. Hence, RC (cutting then reversing) has period four also.

We leave readers with a question: is there a natural move (permutation) that leads to Bottom to Bottom and Period-Four Principles?

References

[1] C. Mulcahy. "Low-Down Triple Dealing," Card Colm, MAA online. http://www.maa.org/community/maa-columns/past-columns-card-colm/low-down-triple-dealing (accessed July 2015).
[2] C. Mulcahy. *Mathematical Card Magic: Fifty-Two New Effects*. A K Peters/CRC Press, Boca Raton, FL, 2013.

11

HEARTLESS POKER

Dominic Lanphier and Laura Taalman

1 Straight, Flush, or Full House?

In standard Five-card poker, some hands are clearly better than others. Even the novice player easily recognizes that Four of a Kind is exciting, while One Pair is comparatively weak. Some players, however, may occasionally need to be reminded which of Full House, Flush, and Straight is the most valuable. Without computing probabilities, it is not immediately obvious which of these three hands is the least likely and thus ranked the highest. The rankings of these hands in a regular poker deck with thirteen ranks (2-10 and J, Q, K, A) and four suits (spades, hearts, clubs, and diamonds) are shown in Figure 11.1.

It is not difficult to compute the necessary probabilities. To build a Full House, you must choose a rank and then three cards in that rank. Then you must choose a second rank, and then two cards in that rank. Since there are thirteen possible ranks and four possible suits (hearts, clubs, spades, and diamonds) in a standard deck of cards, the number of ways to build a Full House is $\binom{13}{1}\binom{4}{3} \cdot \binom{12}{1}\binom{4}{2} = 3{,}744$. If we divide this number of possible Full House hands by the total number $\binom{52}{5} = 2{,}598{,}960$ of possible hands, we see that the probability of being dealt a Full House from a standard deck is $\frac{6}{4{,}165} \approx 0.00144$.

Similar calculations produce the well-known list of the frequencies and probabilities of poker hands shown in Table 11.1. Note that each poker hand is represented exactly once in this table; for example, recall that a *Straight Flush* is a Straight with all cards of the same suit, and a *Royal Flush* is a Straight Flush of 10, J, Q, K, A. Then the number of Flushes excludes those hands that are Straight Flushes, and the number of Straight Flushes excludes Royal Flushes. The last column is the ratio of the frequency of a given poker hand divided by the frequency of the next-higher-ranked hand. For example, the 6 in the Full House row asserts that a Full House is 6 times more likely than a Four of a Kind. Similarly, a player is 17.33 times more likely to get a Four of a Kind than a Straight Flush.

Straight   Flush   Full House

Figure 11.1. A *Straight* is five cards with ranks in sequence. A *Flush* is five cards of the same suit. A *Full House* is three cards of one rank and two cards of another rank. Card images and layout from Encke [2].

TABLE 11.1.
Frequency and probability of hands in standard five-card poker

Poker hand	Calculation	Frequency	Probability	Ratio
Royal Flush	$\binom{4}{1}$	4	0.00000154	—
Straight Flush	$\binom{10}{1}\binom{4}{1} - \binom{4}{1}$	36	0.00001385	9
Four of a Kind	$\binom{13}{1}\binom{4}{4}\cdot\binom{48}{1}$	624	0.00024010	17.333
Full House	$\binom{13}{1}\binom{4}{3}\cdot\binom{12}{1}\binom{4}{2}$	3,744	0.00144058	6
Flush	$\binom{13}{5}\binom{4}{1} - \binom{10}{1}\binom{4}{1}$	5,108	0.00196540	1.364
Straight	$\binom{10}{1}\binom{4}{1}^5 - \binom{10}{1}\binom{4}{1}$	10,200	0.00392465	1.997
Three of a Kind	$\binom{13}{1}\binom{4}{3}\cdot\binom{12}{2}\binom{4}{1}^2$	54,912	0.02112845	5.384
Two Pair	$\binom{13}{2}\binom{4}{2}^2\cdot\binom{11}{1}\binom{4}{1}$	123,552	0.04753902	2.25
One Pair	$\binom{13}{1}\binom{4}{2}\cdot\binom{12}{3}\binom{4}{1}^3$	1,098,240	0.42256902	8.889
High Card	All other hands	1,302,540	0.50117739	1.186

Note: The ratios are approximations.

The common confusion about the rankings of Full House, Flush, and Straight is understandable; being dealt a Flush is just 1.997 times more likely than being dealt a Straight, and being dealt a Full House is just 1.364 times more likely than being dealt a Flush. To see this graphically, consider the logarithmically scaled plot of hand frequencies shown in Figure 11.2. The points on the graph that correspond to the probabilities of obtaining a Full House, Straight, and Flush level off compared to the probabilities of the other hands. This is because the probabilities of obtaining these three hands are relatively close together. This clustering prompts us to focus on these three hands in this chapter. Also, note that the graph shows that the probability of getting One Pair is close to the probability of getting High Card.

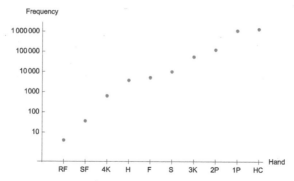

Figure 11.2. Logarithmic plot of poker hand frequencies. Hand abbreviations: RF, royal flush; SF, straight flush; 4K, four of a kind; H, full house; F, flush; S, straight; 3K, three of a kind; 2P, two pair; 1P, one pair; HC, high card.

2 More or Less: Frequencies of Generalized Poker Hands

Poker is typically played with a standard deck of fifty-two cards with thirteen ranks and four suits. There are many variations of poker created by changing various rules, such as how many cards are dealt, how many can be traded, and whether any cards are wild. We could also vary the deck itself, say, by removing certain cards at the outset.

For example, the game of *Heartless Poker* uses a standard deck with all the hearts removed. That there are only three suits and thirty-nine cards in the deck changes the probabilities, and perhaps the rankings, of the different types of hands. With only three suits, we would expect it to be much easier to get a Flush in Heartless Poker than with the usual deck. Moreover, with only three cards of each rank, we might expect it to be more difficult than usual to get a Straight. In fact, in a moment we will see that in Heartless Poker, the rankings of Flush and Straight are reversed from their usual rankings; a Heartless Poker Straight is more valuable than a Heartless Poker Flush, as shown in Figure 11.3.

We could also modify the game of poker by increasing the size of the deck. For example, the commercially available *Fat Pack* deck of cards has the standard thirteen possible ranks, but each rank appears in eight suits (the usual spades, hearts, clubs, and diamonds along with the new suits tridents, roses, hatchets, and doves). With so many cards in each rank, we might expect it to be much easier to get a Straight, and that is in fact the case. Interestingly, as we will see in a moment, it is comparatively easy to get a Full House in Fat Pack Poker, but difficult to get a Flush. In other words, in Fat Pack Poker, the rankings of Full House and Flush are reversed from their usual positions, as illustrated in Figure 11.4.

Heartless Flush   Heartless Straight   Heartless Full House

Figure 11.3. In Heartless Poker, the relative rankings of the Flush and Straight hands are reversed.

Fat Pack Straight   Fat Pack Full House   Fat Pack Flush

Figure 11.4. When playing with the Fat Pack deck, the relative rankings of Flush and Full House hands are reversed.

Basic counting arguments give the poker hand frequencies in Table 11.2 for decks with s suits and r ranks. The entries for Straights and Straight Flushes hold only for $r \geq 6$, since for $r = 5$, the only possible Straight is $A\ 2\ 3\ 4\ 5$ (regardless of whether Ace is considered high or low), and for $r < 5$, it is impossible to construct a Straight hand. Note that the sum of the numbers of hands is $\binom{rs}{5} - \binom{r}{1}\binom{s}{5}$, due to the possibility of obtaining the invalid poker hand Five of a Kind when $s \geq 5$. Note that Takahasi and Futatsuya [3] studied another problem of generalized poker hands: specifically, for a deck with r ranks, s suits, and h cards per hand, they studied the probability of obtaining a High Card, where Straights are allowed to be circular (so K, A, 2, 3, 4 counts as a Straight).

The general frequency formulas in Table 11.2 allow us to prove a small preliminary result.

Theorem 1. *Every possible permutation of Straight, Flush, and Full House rankings occurs for some deck with s suits and r ranks.*

Theorem 1 follows immediately by simply applying the frequency formulas in Table 11.2 to the six (r, s) examples in Table 11.3. The first row is the usual four-suit, thirteen-rank poker deck, the second row is the three-suit, thirteen-

TABLE 11.2.
Frequencies of the possible poker hands for a deck with s suits and r ranks per suit

Poker hand	Number of possible hands	Leading term
Royal Flush	$\binom{s}{1}$	s
Straight Flush	$\binom{r-3}{1}\binom{s}{1} - \binom{s}{1}$	rs
Four of a Kind	$\binom{r}{1}\binom{s}{4}\binom{rs-s}{1}$	$r^2 s^5$
Full House	$\binom{r}{1}\binom{s}{3}\binom{r-1}{1}\binom{s}{2}$	$r^2 s^5$
Flush	$\binom{r}{5}\binom{s}{1} - \binom{r-3}{1}\binom{s}{1}$	$r^5 s$
Straight	$\binom{r-3}{1}\binom{s}{1}^5 - \binom{r-3}{1}\binom{s}{1}$	rs^5
Three of a Kind	$\binom{r}{1}\binom{s}{3}\binom{r-1}{2}\binom{s}{1}^2$	$r^3 s^5$
Two Pair	$\binom{r}{2}\binom{s}{2}^2\binom{r-2}{1}\binom{s}{1}$	$r^3 s^5$
One Pair	$\binom{r}{1}\binom{s}{2}\binom{r-1}{3}\binom{s}{1}^3$	$r^4 s^5$
High Card	$\left[\binom{r}{5} - (r-3)\right]\left[\binom{s}{1}^5 - s\right]$	$r^5 s^5$

Note: The leading terms in the last column are relevant to Section 3.

TABLE 11.3.
Examples of frequencies and rankings of Straight (S), Flush (F), and Full House (H) for decks with s suits and r cards per suit

(r, s)	Straight	Flush	Full House	Ranking
$(13, 4)$	10,200	5,108	3,744	$S < F < H$
$(13, 3)$	2,400	3,831	468	$F < S < H$
$(13, 8)$	327,600	10,216	244,608	$S < H < F$
$(25, 15)$	16,705,920	796,620	28,665,000	$H < S < F$
$(30, 7)$	453,600	997,353	639,450	$F < H < S$
$(33, 9)$	1,771,200	2,135,754	3,193,344	$H < F < S$

Note: Hands with the lowest frequencies are ranked highest.

rank Heartless Poker deck, and the third row is the eight-suit, thirteen-rank Fat Pack deck.

3 Are Straight, Flush, and Full House Ever Tied?

We now come to the main question of this chapter: are there any generalized poker decks for which the Straight, Flush, and Full House hands have the same frequency, either pairwise or all together? We handle the latter part of the question in this section and the pairwise question in the next.

From the leading terms in the last column of Table 11.2, we see that the relative rankings of Full House and Flush will switch infinitely often as r grows and then as s grows, and so on. However, this is not the case for the relative rankings of Full House and Straight; although Full House is ranked higher than Straight in standard poker, we will see that this occurs only in a finite number of cases. To aid in our search for pairs (r, s) of rank and suit sizes for which there are ties between hands, we investigate three curves and their intersections.

Definition 1. For $r \geq 6$ and $s \geq 1$, define the following three curves:

$$C_{SF}(r, s): \quad r(r - 1)(r - 2)(r - 4) = 120s^4,$$

$$C_{SH}(r, s): \quad r(r - 1)s(s - 1)(s - 2) = 12(r - 3)(s + 1)(s^2 + 1),$$

$$C_{FH}(r, s): \quad r(r - 1)(r - 2)(r - 3)(r - 4) = 120(r - 3)$$
$$+ 10r(r - 1)s(s - 1)^2(s - 2).$$

The equation $C_{SF}(r, s)$ is obtained by setting the frequencies of Straight (S) and Flush (F) from Table 2 equal to each other and simplifying. This means that a point (r, s) is on the curve $C_{SF}(r, s)$ if and only if the Straight and Flush hands have equal frequency in a generalized poker deck with r ranks and s suits. Similarly, C_{SH} describes the decks for which Straight and Full House (H) have equal frequency, and C_{FH} describes the decks for which Flush and Full House have equal frequency.

Figure 11.5 shows the graphs of $C_{SF}(r, s)$, $C_{SH}(r, s)$, and $C_{FH}(r, s)$ for $6 \leq r \leq 50, 1 \leq s \leq 20$. The decreasing curve is $C_{SH}(r, s)$, the steeper of the two increasing curves is $C_{SF}(r, s)$, and the remaining curve is $C_{FH}(r, s)$. The regions bounded by the curves represent where one poker hand outranks another. For example, the six dots in the figure represent the six examples in Table 11.3.

The point in Figure 11.5 at which the three curves meet represents a value (r, s) for which the frequency formulas for S, F, and H are all equal. In other words, for that value of (r, s), the probabilities of getting a Straight, Flush, and Full House are all equal. Our first result is to show that there is only one such point, and that it has noninteger coordinates.

Theorem 2. *If $r \geq 6$ and $s \geq 1$, then there exists exactly one point (r, s) that simultaneously satisfies $C_{SF}(r, s)$, $C_{SH}(r, s)$, and $C_{FH}(r, s)$, and this solution has $22 < r < 23$ and $6 < s < 7$. Consequently, there is no generalized poker deck with r ranks and s suits for which Straight, Flush, and Full House together have equal ranking.*

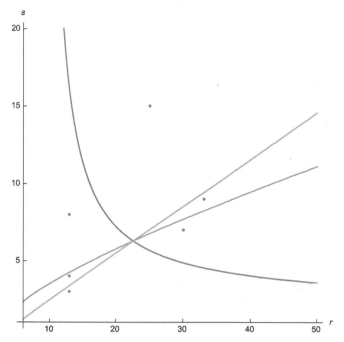

Figure 11.5. Curves $C_{SF}(r, s)$ (green), $C_{SH}(r, s)$ (blue), and $C_{FH}(r, s)$ (yellow) intersect at exactly one point.

Proof. Recall that a point on $C_{SF}(r, s)$ is where the probability of getting a Straight is equal to the probability of getting a Flush, and a point on $C_{SH}(r, s)$ is where the probability of getting a Straight is equal to the probability of getting a Full House. So any point on both curves must have the frequency of Straight equal to the frequency of a Flush and a Full House. Thus, the frequency of all three hands must be the same. So it suffices to show that there is only one (r, s) that simultaneously satisfies $C_{SH}(r, s)$ and $C_{SF}(r, s)$. Implicitly differentiating $C_{SF}(r, s)$ gives

$$480s^3 \frac{ds}{dr} = \frac{df}{dr},$$

where $f(r) = r(r - 1)(r - 2)(r - 4)$. Since it is clear that $\frac{df}{dr} > 0$ for $r \geq 6$, and since $s \geq 1$ by hypothesis, we have $\frac{ds}{dr} > 0$ for points (r, s) on C_{SF}.

For $g(r) = \frac{r(r-1)}{12(r-3)}$, we have $g'(r) = \frac{r^2-6r+3}{12(r-3)^2} > 0$ for $r \geq 6$. Thus $g(r)$ is increasing for such r, and so $g(r) \leq g(20) \approx 1.86$ for $6 \leq r \leq 20$. Now, for

(r, s) on $C_{SH}(r, s)$, we have

$$g(r) = h(s) = \frac{(s + 1)(s^2 + 1)}{s(s - 1)(s - 2)},$$

and so $h(s) \leq g(20) < 1.863$ for such (r, s) with $6 \leq r \leq 20$. A straightforward calculation shows that $h(s) \leq 1.863$ implies

$$6.589s^2 + 1 \leq 0.863s^3 + 2.726s.$$

This inequality clearly does *not* hold for $s = 1$ and $s = 6$, and these two polynomials are increasing and concave up for $s \geq 1$. Now, the derivative of $0.863s^3 + 2.726s$ is smaller than the derivative of $6.589s^2 + 1$ for $1 \leq s \leq 4.874$ and larger for $4.874 \leq s$. Furthermore, we have

$$6.589s^2 + 1 > 0.863s^3 + 2.726s$$

at $s = 4.874$. Therefore, as

$$6.589s^2 + 1 > 0.863s^3 + 2.726s$$

at $s = 6$ also, then it must hold for all $1 \leq s \leq 6$. It follows that $h(s) > 1.863$ for $s \leq 6$. So we must have $s > 6$ for $r \leq 20$. However, if a point (r, s) with $r \leq 20$ is also on the curve $C_{SF}(r, s)$ then we must have

$$s = \left(\frac{r(r - 1)(r - 2)(r - 4)}{120}\right)^{1/4} \leq \left(\frac{20 \cdot 19 \cdot 18 \cdot 16}{120}\right)^{1/4} \approx 5.49.$$

So we must have $r \geq 20$. From the same argument in the previous line, it follows that we must also have $s \geq 5.49$.

Implicitly differentiating $C_{SH}(r, s)$ shows that ds/dr is equal to

$$\frac{12(s + 1)(s^2 + 1) - (2r - 1)s(s - 1)(s - 2)}{\begin{bmatrix} r(r - 1)(s - 1)(s - 2) + r(r - 1)s(s - 2) + r(r - 1)s(s - 1) \\ -12(r - 3)(s^2 + 1) - 24(r - 3)(s + 1)s \end{bmatrix}}.$$

For $s \geq 5$, we have $3(s + 1) < \frac{39}{6}(s - 2)$ and $4(s^2 + 1) < 6s(s + 1)$. It follows that the numerator of ds/dr is less than 0 for $s \geq 5$. For the denominator, if $j(s) = \frac{3s^2 + 2s + 1}{3s^2 - 6s + 2}$, then $j'(s) = \frac{-24s^2 + 6s + 8}{(3s^2 - 6s + 2)^2} < 0$ for $s \geq 5$. Thus $j(s)$ is decreasing and so $j(s) \leq j(5) = 1.83$ for $s \geq 5$. Recall that $g(r)$ from the previous paragraph is at least 1.86 for $r \geq 20$ and so we must have $g(r) > j(s)$ for $r \geq 20$ and $s \geq 5$. Thus we have

$$\frac{r(r - 1)}{12(r - 3)} > \frac{3s^2 + 2s + 1}{3s^2 - 6s + 2}$$

and, rewriting, this is

$$r(r-1)\left((s-1)(s-2) + s(s-2) + s(s-1)\right) > 12(r-3)\left((s^2 + 1) + 2s(s + 1)\right).$$

Thus the denominator of ds/df is positive for such r and s and so $\frac{ds}{dr} < 0$ for points (r, s) on $C_{SH}(r, s)$.

Therefore $C_{SF}(r, s)$ and $C_{SH}(r, s)$ have at most one intersection point, and this clearly occurs at the point indicated in Figure 11.5. $\qquad\square$

4 Never a Tie: Applications of Diophantine Equations to Poker

Our main result is that Straight, Flush, and Full House have distinct frequencies in any generalized poker game. The proof primarily uses elementary number theory and some elementary analysis. However, a fairly deep result in Diophantine equations is also necessary. The following theorem was shown in Ljunggren [4] and generalized in Bennett and DeWeger [1].

Theorem 3 (Bennett/DeWeger/Ljunggren). *Let n, m be integers. The equation*

$$|nx^4 - my^4| = 1$$

has at most one solution in positive integers x and y.

Results about Diophantine equations often seem to have few applications outside of number theory. Nevertheless, Theorem 3 plays a crucial role in the proof of the following result.

Theorem 4. *There is no pair of integers (r, s) with $r \geq 6$ and $s \geq 1$ that satisfies any of $C_{SF}(r, s)$, $C_{SH}(r, s)$, and $C_{FH}(r, s)$. Consequently, there is no generalized poker deck with r ranks and s suits for which any pair of Straight, Flush, or Full House have the same ranking.*

Proof. We show that each of the three equations has no solution with integers $r \geq 6$ and $s \geq 1$, starting with $C_{FH}(r, s)$. Suppose that (r, s) is an integral solution of $C_{FH}(r, s)$, and consider the equation $C_{FH}(r, s)$ modulo $r - 1$. This gives $0 \equiv -240 \pmod{r - 1}$, which implies that $r - 1$ divides $240 = 2^4 \cdot 3 \cdot 5$. In contrast, considering equation $C_{FH}(r, s)$ modulo r, we have $0 \equiv -360 \pmod{r}$, which implies that r divides $360 = 2^3 \cdot 3^2 \cdot 5$. Since r and $r - 1$ are relatively prime, this means that $r(r - 1)$ divides $2^4 \cdot 3^2 \cdot 5 = 720$. Therefore $r(r - 1) \leq 720$, and so we have

$$r^2 - r - 720 \leq 0.$$

By the quadratic formula, we have $r \leq 27$. The only integers $6 \leq r \leq 27$ for which $r \mid 360$ and $(r - 1) \mid 240$ are $r = 6$ and $r = 9$.

For $r = 6$, equation $C_{FH}(r, s)$ becomes

$$760 = 360 + 300s(s - 1)^2(s - 2),$$

which gives $6/5 = s(s - 1)^2(s - 2)$. This equation that has no solution for integral s. For $r = 9$, equation $C_{FH}(r, s)$ is

$$15{,}120 = 720 + 720s(s - 1)^2(s - 2),$$

which simplifies to $20 = s(s - 1)^2(s - 2)$. Clearly, a positive integral solution to this equation must satisfy $s > 2$. Also, $f(s) = s(s - 1)^2(s - 2)$ is increasing for $s > 2$, and since $f(2) = 0$, $f(3) = 12$, and $f(4) = 72$, there is no integer s so that $f(s) = 20$.

We now turn our attention to $C_{SH}(r, s)$. We show that there are only a finite number of positive integral pairs (r, s) for which a Full House outranks a Straight. In particular, we show that

$$r(r - 1)s(s - 1)(s - 2) > 12(r - 3)(s + 1)(s^2 + 1)$$

for $r \geq 23$ and $s \geq 7$, and then later investigate the cases for smaller r and s. First note that for $r \geq 24$, we have $r(r - 1) > r(r - 3) \geq 24(r - 3)$. It now suffices to prove that $2s(s - 1)(s - 2) > (s + 1)(s^2 + 1)$, or equivalently, that

$$\frac{(s + 1)(s^2 + 1)}{s(s - 1)(s - 2)} = 1 + \frac{4s^2 - s + 1}{s^3 - 3s^2 + 2s} < 2.$$

We have $s^3 - 3s^2 + 2s < 4s^2 - s + 1$ exactly when $g(s) = s^3 - 7s^2 + 3s - 1 > 0$. This follows for $s \geq 7$, because $g'(s)$ is positive for $s \geq 5$ and $g(7) = 20 > 0$.

To prove that $C_{SH}(r, s)$ has no integral solutions, it now suffices to show that the equation does not hold for the cases $s \in \{1, \ldots, 6\}$ and $r \in \{6, \ldots, 22\}$. The cases $s = 1, 2$ are obvious. For $s = 3$, the equation C_{SH} becomes

$$r(r - 1)6 = 80(r - 3),$$

or equivalently, $r^2 - 81r + 240 = 0$, which has no integral solutions. The cases $s = 4, 5, 6$ are similar. Note that for $r = 6$, the equation C_{SH} becomes

$$5s(s - 1)(s - 2) = 6(s + 1)(s^2 + 1),$$

or equivalently, $s^3 + 21s^2 - 4s + 6 = 0$, which has no integral solutions for any s, because $s^3 + 4s^2 - 4s + 6 > 0$ for all s. The cases $r = 7, \ldots, 22$ can be done similarly.

Finally, we must prove that the equation $C_{SF}(r, s)$ given by

$$r(r - 1)(r - 2)(r - 4) = 120s^4$$

has no integral solutions with $r \geq 6$ and $s \geq 1$ (i.e., that Straight and Flush can never be tied in any generalized (r, s) poker game). Here we employ the result in Diophantine equations (Theorem 3). If (r, s) is a pair of positive integers with $r(r - 1)(r - 2)(r - 4) = 120s^4$, then we can write

$$r = \alpha u^4, \ r - 1 = \beta v^4, \ r - 2 = \gamma z^4, \text{ and } r - 4 = \delta w^4,$$

where $\alpha, \beta, \gamma, \delta, u, v, z, w$ are positive integers, and $\alpha\beta\gamma\delta = 120 = 2^3 \cdot 3 \cdot 5$. This gives rise to the following set of Diophantine equations:

$$\alpha u^4 - \beta v^4 = 1,$$

$$\beta v^4 - \gamma z^4 = 1,$$

$$\alpha u^4 - \gamma z^4 = 2,$$

$$\gamma z^4 - \delta w^4 = 2,$$

$$\beta v^4 - \delta w^4 = 3,$$

$$\alpha u^4 - \delta w^4 = 4.$$

We show that no integer α dividing $2^3 \cdot 3 \cdot 5$ and satisfying all the above conditions can exist.

First note that if $2^3 \mid \alpha$, then the third of our Diophantine equations implies that γz^4 is even. Since $\alpha\gamma \mid 2^3 \cdot 3 \cdot 5$, this means that γ is odd; therefore z^4 is even, and thus z is even. Then $z^4 = 2^4 z'^4$ for some $z' \in \mathbb{Z}$. Therefore the third equation implies that

$$2 = \alpha u^4 - \gamma z^4 = 2^3 \alpha' u^4 - 2^4 \gamma z'^4$$

and so $1 = 2^2 \alpha' u^4 - 2^3 \gamma z'^4$, which gives a contradiction, since the right-hand side of the equation is even. Therefore it follows that $2^3 \nmid \alpha$.

In a similar way, if $2^2 \mid \alpha$, then it follows that γ must be even and hence β and δ are odd.

To tackle the rest of the cases, we repeatedly apply a couple of ingredients. The first is the observation that for any $a \in \mathbb{Z}$, we have

$$a^4 \equiv 1 \text{ or } 0 \pmod 3 \text{ and } a^4 \equiv 1 \text{ or } 0 \pmod 5. \tag{1}$$

The second ingredient is Theorem 3, which is a classical result in Diophantine equations.

Suppose that $3 \mid \alpha$. Then we have

$$\alpha \in \{3, 2 \cdot 3, 2^2 \cdot 3, 3 \cdot 5, 2 \cdot 3 \cdot 5, 2^2 \cdot 3 \cdot 5\}.$$

Taking the first of the equations modulo 3, we get $\alpha u^4 - \beta v^4 \equiv 1 \pmod 3$ and thus $-\beta v^4 \equiv 1 \pmod 3$. Thus we have $-\beta \equiv 1 \pmod 3$ and so $\beta \equiv 2 \pmod 3$. We can do the same thing to the equations

$$\alpha u^4 - \delta w^4 = 4 \quad \text{and} \quad \alpha u^4 - \gamma z^4 = 2.$$

We get $\delta \equiv 2$ (mod 3) and $\gamma \equiv 1$ (mod 3). Note also that $\beta v^4 - \delta w^4 \equiv 3$ (mod 5). Combining this with the fact that $\alpha \beta \gamma \delta = 2^3 \cdot 3 \cdot 5$, we see that the ordered quadruple of integers $(\alpha, \beta, \gamma, \delta)$ can only be one of

$$(3, 2^3, 1, 5), \quad (3, 2^2 \cdot 5, 1, 2), \quad (2^2 \cdot 3, 5, 1, 2), \quad (3, 5, 2^2, 2). \tag{2}$$

From equations (1), the first of these quadruples implies that $z^4 - 5w^4 = 2$. Taking this modulo 5 gives $z^4 \equiv 2$ (mod 5), which is a contradiction. The second quadruple gives $3u^4 - 2^2 \cdot 5v^4 \equiv 1$ (mod 5), which implies $3u^4 \equiv 1$ (mod 5). So $u^4 \equiv 2$ (mod 5), which again gives a contradiction. We previously showed that if $2^2 \mid \alpha$, then γ must be even, so we can immediately eliminate the third quadruple.

Theorem 3 implies that the equation

$$\beta v^4 - \gamma z^4 = 1$$

has at most one solution for v, z positive integers. For the fourth quadruple in the list (2), $\beta = 5$ and $\gamma = 2^2$, and we have the solution $v = z = 1$. By Theorem 3, there can be no more solutions with v, z positive integers. Now, $v = z = 1, \beta = 5, \gamma = 2^2$ together imply $r = 6$, and then equation $C_{SF}(r, s)$ simplifies to $s^4 = 2$, which has no integral solutions. Therefore, we cannot have this case, and this eliminates the fourth quadruple in list (2). Thus we have that $3 \nmid \alpha$.

Now suppose that $5 \mid \alpha$. Then

$$\alpha u^4 - \beta v^4 \equiv 1 \quad (\text{mod } 5)$$

means that $-\beta v^4 \equiv 1$ (mod 5), and from condition (1), we have that $\beta \equiv 4$ (mod 5). Therefore, the only possibilities for β are 4 and 24. Note that

$$\alpha u^4 - \beta v^4 = 1$$

implies $(\alpha, \beta) = 1$, and as $3 \nmid \alpha$ we must have $\alpha = 5$. If $\beta = 4$ then

$$|\alpha u^4 - \beta v^4| = 1$$

has the solution $u = v = 1$, and by Theorem 3, there are no other solutions with u, v positive integers. Thus we have $r = \alpha u^4 = 5$ and equation $C_{SF}(r, s)$ simplifies to $2s^4 = 1$, which has no integral solutions. Therefore $\beta \neq 4$, and thus $\beta = 24$ is the only possible value for β. Thus the only possible values of the constants are given by the quadruple $(\alpha, \beta, \gamma, \delta) = (5, 2^3 \cdot 3, 1, 1)$. Then the fourth of our Diophantine equations becomes

$$z^4 - w^4 = 2.$$

But by condition (1), we know that z^4, w^4 can only be 0 or 1 modulo 5, and so $z^4 - w^4$ can only be 0, 1, or 4 modulo 5. Hence we must have that $5 \nmid \alpha$.

We are now left with the cases where α is one of 1, 2, or 2^2. If $\alpha = 2^2$, then, as mentioned previously, γ is even and β, δ are odd. If $|\alpha - \beta| = 1$, then by Theorem 3, we must have $u = v = 1$ and so $r = \alpha u^4 = 4$. As $r \geq 5$, this cannot occur, and it follows that $\beta \neq 3, 5$. Thus, the only possibilities for β are 1 and 15. If $\beta = 15$, then $\alpha u^4 - \beta v^4 \equiv 1 \pmod 5$ gives $4u^4 \equiv 1 \pmod 5$, which again contradicts condition (1). So we must have $\beta = 1$; but now we have

$$\alpha u^4 - \beta v^4 = 1 = 4u^4 - v^4 = (2u^2 - v^2)(2u^2 + v^2),$$

or equivalently, $2u^2 + v^2 = 1$, which has only the solution $u = 0$, $v^2 = 1$. This implies that $r = 0$ and thus gives a contradiction. Therefore $\alpha \neq 2^2$.

Now for the case $\alpha = 2$. If $3 \mid \beta$, then consider the equation $2u^4 - \beta v^4 = 1$ (mod 3). We get $2u^4 \equiv 1 \pmod 3$ and so $u^4 \equiv 2 \pmod 3$, which gives a contradiction. Thus $3 \nmid \beta$ and in a similar way, by taking the equation modulo 5, we have that $5 \nmid \beta$. We can do the same thing for δ using the equation $2u^4 - \delta w^4 = 4$ and get $3, 5 \nmid \delta$. Thus we must have that $3 \cdot 5 \mid \gamma$. From the second Diophantine equation $\beta v^4 - \gamma z^4 = 1$, we get that $\beta v^4 \equiv 1 \pmod 3$ and $\beta v^4 \equiv 1 \pmod 5$. Thus we have $\beta \equiv 1 \pmod 3$ and $\beta \equiv 1 \pmod 5$. But as $3, 5 \nmid \beta$, then β can only be 1, 2, or 2^2, and only $\beta = 1$ is congruent to 1 modulo 5. But then the equation $\alpha u^4 - \beta v^4 = 1$ becomes $2u^4 - v^4 = 1$, which by Theorem 3 only has one solution, namely, $u = v = 1$. This implies $r = 2$, but by hypothesis, we have $r \geq 5$. So we must have $\alpha \neq 2$.

Finally, if $\alpha = 1$, then consider the equation $u^4 - \gamma z^4 = 2$. Since $u^4 \equiv 0$ or 1 (mod 3 or 5), we have that $\gamma z^4 \equiv 1$ or 2 (mod 3) and $\gamma z^4 \equiv 3$ or 4 (mod 5). It follows that $\gamma \equiv 1$ or 2 (mod 3) and $\gamma \equiv 3$ or 4 (mod 5). Thus, the only possibilities for γ are 2^2 or 2^3, and so we have $3 \cdot 5 \mid \beta \delta$. If $5 \mid \delta$, then we take the equation $\gamma z^4 - \delta w^4 = 2 \pmod 5$ and we get $\gamma \equiv 2 \pmod 5$. This cannot hold for $\gamma = 2^2$ or 2^3, so we must have $5 \nmid \delta$ and so $5 \mid \beta$ must occur. Taking $\beta v^4 - \gamma z^4 = 1 \pmod 5$, we get that $\gamma \equiv 4 \pmod 5$, and so $\gamma = 2^2$ is the only possibility. Because $\alpha = 1$, $\gamma = 2^2$, and $5 \mid \beta$, the only possibilities for δ are 1, 3, or 6. However, taking the equation $\beta v^4 - \delta w^4 = 3 \pmod 5$, we get that $\delta \equiv 2 \pmod 5$. Therefore, none of the possibilities work, and thus $\alpha \neq 1$.

This exhausts all possibilities for α. Thus there can be no integral solutions for the curve $C_{SF}(r, s)$. $\qquad\qquad\qquad\qquad\qquad\qquad\qquad\qquad\qquad\qquad\quad\square$

References

[1] M. A. Bennett and B.M.M. DeWeger. On the Diophantine equation $|ax^n - by^n| = 1$. *Math. Comp.* **67** no. 221 (1998) 413–438.

[2] O. Encke. LaTeX file poker.sty, copyright 2007–2008. www.encke.net (accessed November 20, 2012).

[3] K. Takahasi and M. Futatsuya. On a problem of a probability arising from poker. *J. Japan Statist. Soc.* **14** no. 1 (1984) 11–18.

[4] W. Ljunggren. Einige Eigenschaften der Einheitenreeler quadratischer und rein biquadratischer Zahlkörper mit Anwendung auf die Lösung einer Klasse von bestimmter Gleichungen vienten Grades. *Det Norske Vidensk. Akad. Oslo Skuifter* *I* **12** (1936) 1–73.

12

AN INTRODUCTION TO GILBREATH NUMBERS

Robert W. Vallin

In 1958, an undergraduate math major at the University of California, Los Angeles, named Norman Gilbreath published a note in *The Linking Ring*, the official publication of the International Brotherhood of Magicians [4], in which he described a certain card trick. Stated succinctly, this trick can be performed by handing an audience member a deck of cards, letting him or her cut the deck several times, and then having him or her deal N cards from the top into a pile. The audience member takes the two piles (the cards in hand and the set now piled on the table) and riffle-shuffles them together. The magician now hides the deck (under a cloth, perhaps, or behind the back) and proceeds to produce one pair of cards after another in which one card is black and the other is red.

The key to this trick is that the cards are prearranged in black/red order before the deck is handed out. Cutting the deck does not change this arrangement. When the top N cards are dealt into a pile, the black/red parity is still there, but the order of the cards in the packet is reversed. It then takes a mathematical induction argument to show that however the riffle-shuffle is performed, consecutive pairs of cards still maintain opposite colors. Let us look at a small example with eight cards (Figure 12.1). The subscripts refer to the first, second, third, etc. card in the original deck with R for red and B for black. The process is illustrated in Figure 12.1.

If cards are taken two at a time from either the top or bottom of the deck, then we get pairs consisting of one of each color.

It is interesting to note that having a pattern of *two* types of characteristics does not matter. This trick also works if the cards are arranged by suit (e.g., Clubs, Hearts, Spades, and Diamonds) and, after shuffling, are dealt off four at a time.

Let us analyze this behavior mathematically. For any nonempty set S, a permutation on S is a bijection from S onto itself. We are only concerned with permutations on sets of numbers, beginning with the finite sets $\{1, 2, 3, \ldots, N\}$, and eventually moving on to the natural numbers generally, with a particular property (described below). Denote our permutation as π,

Original order	After cutting	Two piles	Shuffled
B_1	B_5	B_1	R_8
R_2	R_6	R_2	B_7
B_3	B_7	B_3	B_1
R_4	R_8	R_4	R_6
B_5	B_1		R_2
R_6	R_2	R_8	B_5
B_7	B_3	B_7	B_3
R_8	R_4	R_6	R_4
		B_5	

Figure 12.1. Eight-card version of Gilbreath's trick.

Original order	Two "piles"	"Shuffled"
1	5	5
2	6	4
3	7	6
4	8	7
5	9	3
6	10	8
7		2
8	4	9
9	3	1
10	2	10
	1	

Figure 12.2. Example of a Gilbreath permutation.

and for any j, $\pi(j)$ refers to the number in the jth place of the permutation (and not the placement of the number j). So, if we permute $\{1, 2, 3, 4, 5\}$ to get $\{3, 4, 5, 2, 1\}$, then we have $\pi(1) = 3$ and $\pi(5) = 1$.

Let us repeat the process shown in Figure 12.1, but this time with numbers rather than cards (Figure 12.2). Now that we are not performing a trick, the cutting (which gives the audience the false idea of randomness) is unnecessary.

We can express the result of shuffling as a permutation on $\{1, 2, 3, \ldots, N\}$. In the example shown in Figure 12.2, we obtain

$$\{5, 4, 6, 7, 3, 8, 2, 9, 1, 10\}.$$

A permutation obtained in this way is known as a *Gilbreath permutation*.

Not every permutation is a Gilbreath permutation. In fact, while there are $N!$ permutations on $\{1, 2, 3, \ldots, N\}$, there are only 2^{N-1} Gilbreath permutations. This fact, and the *Ultimate Gilbreath Principle* below (which tells us which permutations are Gilbreath permutations), are proved in Diaconis and Graham [2].

Theorem 1 (The Ultimate Gilbreath Principle). *The following are equivalent, for any permutation π of $\{1, 2, 3, \ldots, N\}$.*

1. *π is a Gilbreath permutation.*
2. *For each j, the first j values*

$$\{\pi(1), \pi(2), \ldots, \pi(j)\}$$

   are distinct modulo j.
3. *For each j and k with $jk \leq N$ the values*

$$\{\pi((k-1)j+1), \pi((k-1)j+2), \ldots, \pi(kj)\}$$

   are distinct modulo j.
4. *For each j, the first j values comprise a (possibly unordered) set of consecutive numbers from among $1, 2, 3, \ldots, N$.*

Property three makes the magic trick work. It implies that whether you are taking cards from the deck in groups of two (red/black) or four (Clubs, Hearts, Spades, Diamonds), if the deck is initially set up correctly, then the magician gets one card of each type. Our present interest is in the fourth property. It implies that even if the j numbers are not currently written in order, they appear in order in $\{1, 2, 3, \ldots, N\}$. So, for example, the permutations $\{2, 3, 1, 4\}$ and $\{3, 4, 2, 1\}$ are Gilbreath permutations of $\{1, 2, 3, 4\}$, while $\{3, 4, 1, 2\}$ is not.

1 Continued Fractions

We are interested in expressing Gilbreath permutations in the form of continued fractions. For that reason, we discuss some basic properties of continued fractions now.

Definition 1. Let a_i, $i = 0, 1, 2, 3, \ldots$ denote a collection (possibly finite) of positive integers.[1] A *simple continued fraction* is an expression of the form

$$a_0 + \cfrac{1}{a_1 + \cfrac{1}{a_2 + \cfrac{1}{a_3 + \ddots}}}$$

To simplify our notation, we express this continued fraction as

$$[a_0; a_1, a_2, a_3, \ldots].$$

---

[1] Some references use complex numbers with integer coefficients.

The individual $a_0, a_1, a_2, \ldots$ are referred to as the *partial quotients* of the continued fraction expansion.

A good general reference on continued fractions is the book by Olds [5]. In what follows, we are only concerned with expressing numbers in the unit interval.

Example 1. *The finite, simple continued fraction* $[2; 3, 6]$ *represents the number*

$$2 + \cfrac{1}{3 + \frac{1}{6}} = 2 + \frac{1}{\frac{19}{6}} = 2 + \frac{6}{19} = \frac{44}{19}.$$

In the other direction, we can take, say, the fraction $\frac{32}{15}$ *and write*

$$\frac{32}{15} = 2 + \frac{2}{15} = 2 + \frac{1}{\frac{15}{2}} = 2 + \cfrac{1}{7 + \frac{1}{2}}.$$

So $32/15 = [2; 7, 2]$.

The technique used in Example 1 for representing a rational number as a continued fraction is guaranteed to terminate. Notice that in this dividing process ($32 \div 15$, then $15 \div 2$), the remainders, which in the next step become the denominators, are strictly decreasing. Thus, the remainder must at some point become the number 1, finishing our continued fraction expansion in the rational numbers.

If x is a positive irrational number, then there exists a largest integer a_0 such that $x = a_0 + \frac{1}{x_1}$ with $0 < \frac{1}{x_1} < 1$. Note that we have

$$x_1 = \frac{1}{x - a_0} > 1,$$

which is irrational (since x is irrational and a_0 is an integer).

Repeat this process, starting with x_1. We find a_1, the greatest integer such that

$$x_1 = a_1 + \frac{1}{x_2},$$

where $0 < \frac{1}{x_2} < 1$. As we continue, we generate the continued fraction for x:

$$[a_0; a_1, a_2, a_3, \ldots],$$

where this time the sequence a_i does not terminate.

Example 2. *We can express* $\sqrt{3}$ *as*

$$\sqrt{3} = [1; 1, 2, 1, 2, 1, 2, \ldots] \text{ or, more conveniently, } [1; \overline{1, 2}].$$

This continued fraction expansion can be verified by solving the equation

$$x = 1 + \cfrac{1}{1 + \cfrac{1}{2 + (x - 1)}},$$

where $(x - 1)$ is the repeating part of our continued fraction.

A quadratic irrational is a number of the form

$$\frac{P \pm \sqrt{D}}{Q},$$

where P, Q, and D are integers, $Q \neq 0$, and D is a non-square satisfying $D > 1$. In 1779, Lagrange proved that the continued fraction expansion of any quadratic irrational will eventually become periodic. Later, Euler showed the converse. If a continued fraction is eventually periodic, then the value can be expressed as a quadratic irrational number.

There are also nonrepeating, infinite, simple continued fractions. An example is

$$\pi = [3; 7, 15, 1, 292, 1, 1, \ldots].$$

We know this fraction cannot end or repeat, since π is a transcendental number (i.e., not the solution to a polynomial with integer coefficients).

We now turn our attention to *convergents*.

Definition 2. Let x have the simple continued fraction expansion (finite or infinite) $[a_0; a_1, a_2, a_3, \ldots]$. The *convergents* are the sequence of finite simple continued fractions

$$c_0 = a_0, \qquad c_1 = a_0 + \frac{1}{a_1}, \qquad c_2 = a_0 + \cfrac{1}{a_1 + \frac{1}{a_2}}, \qquad \ldots.$$

In general, $c_n = [a_0; a_1, a_2, a_3, \ldots, a_n]$.

Typically, the next step is to represent these convergents in the form of rational numbers. For example, we can write

$$c_0 = a_0 = \frac{p_0}{q_0},$$

$$c_1 = a_0 + \frac{1}{a_1} = \frac{p_1}{q_1},$$

$$c_2 = a_0 + \cfrac{1}{a_1 + \frac{1}{a_2}} = \frac{p_2}{q_2},$$

and so on.

These convergents approach a limit. If we start with two convergents c_i and c_{i+1}, it is well known that

$$c_{i+1} - c_i = \frac{p_{i+1}}{q_{i+1}} - \frac{p_i}{q_i} = \frac{p_{i+1}q_i - p_i q_{i+1}}{q_{i+1}q_i} = \frac{(-1)^{i+1}}{q_{i+1}q_i}.$$

From the definition of convergent, it can be seen that the sequence q_i is always positive and increasing. Thus we have the following result.

Theorem 2. *For any simple continued fraction $[a_0; a_1, a_2, a_3, \ldots]$ the convergents c_i form a sequence of real numbers, where for all i, we have*

- $c_{2i-1} < c_{2i+1} < c_{2i}$

  and

- $c_{2i+1} < c_{2i+2} < c_{2i}.$

Thus, the sequence of convergents has the property

$$c_1 < c_3 < c_5 < \cdots < c_{2i-1} < \cdots < c_{2i} < \cdots < c_4 < c_2 < c_0.$$

We also need the following theorem.

Theorem 3. *Let $[a_0; a_1, a_2, a_3, \ldots]$ represent an infinite, simple continued fraction. Then there is a point x on the real line such that $x = [a_0; a_1, a_2, a_3, \ldots]$.*

2 Gilbreath Continued Fractions

We now wish to take a Gilbreath permutation, such as $\{3, 4, 2, 1\}$, and turn it into a continued fraction. This fraction then will represent a real number in the unit interval. We accomplish this by taking a permutation, say $\{3, 4, 2, 1\}$, and writing it as

$$0 + \cfrac{1}{3 + \cfrac{1}{4 + \cfrac{1}{2 + \cfrac{1}{1}}}}.$$

It is easily checked that this is equal to $\frac{13}{42}$. Let us refer to continued fractions produced in this way as *Gilbreath continued fractions*. Rather than write out the full continued fraction, we can take advantage of the fact that all the numerators are 1. Therefore, we can write the number in bracket notation, noting that the whole-number part is 0:

$$0 + \cfrac{1}{3 + \cfrac{1}{4 + \cfrac{1}{2 + \cfrac{1}{1}}}} = [0; 3, 4, 2, 1].$$

In working with continued fractions, it is customary not to allow a representation to end with a 1. That is because $[0; 3, 4, 2, 1]$ is the same as $[0; 3, 4, 3]$. If the terminal digit is not allowed to be 1, then continued fraction representations are unique. This is not the case with decimal representations, where $1.000\ldots = 0.999\ldots$. This is one of the perquisites in dealing with continued fractions rather than decimals. However, following this convention creates issues with Gilbreath permutations, so we break with tradition by allowing the expansion to terminate with a 1.

If we begin with a Gilbreath permutation of length N,

$$\{\pi(1), \pi(2), \ldots, \pi(N)\},$$

we can extend it to a permutation of $\mathbb{N}$ by inserting the numbers $N + 1$, $N + 2$, $N + 3, \ldots$, in order, at the end. In this way, we preserve the property of being a Gilbreath permutation. Note that, as a consequence of property 4 of Theorem 1, inserting these numbers in any other way would produce something that is not a Gilbreath permutation.

For example, we can extend $\{4, 3, 5, 2, 1, 6\}$, producing the sequence

$$\{4, 3, 5, 2, 1, 6, 7\}, \{4, 3, 5, 2, 1, 6, 7, 8\}, \{4, 3, 5, 2, 1, 6, 7, 8, 9\}, \ldots$$

and eventually obtaining the infinite ordered set

$$\{4, 3, 5, 2, 1, 6, 7, 8, 9, 10, 11, 12, 13, 14, \ldots\}.$$

The point in the expansion beyond which we have $a_k = k$ is referred to as the place where the permutation *straightens out*.

Of course, each of these finite strings can be the terms in a finite simple continued fraction (representing a rational number). The infinite sequence of numbers converges to an irrational number equivalent to the infinite simple continued fraction

$$[0; 4, 3, 5, 2, 1, 6, 7, 8, 9, 10, 11, 12, \ldots].$$

3 An Analytic Perspective on Gilbreath Continued Fractions

We refer to our continued fractions using the notation $[0; a_1, a_2, a_3, \ldots, a_n]$, recalling that $a_j = \pi(j)$. Let $\mathcal{G}$ represent the set of finite and infinite Gilbreath continued fractions. Moreover, let $\mathcal{G}_F$ denote the set of Gilbreath numbers whose representation is a finite string. Thus, $\mathcal{G}_F$ denotes the set of rational numbers in $\mathcal{G}$. Likewise, let $\mathcal{G}_I$ denote the set of Gilbreath numbers whose representation is an infinite string, which is to say that they are the irrationals in $\mathcal{G}$. Together, these form a subset of the unit interval $[0, 1]$, and the natural question to ask is "How much of the unit interval is taken up by these numbers?" We now show that $\mathcal{G}$ is a countably infinite set and a very sparse set with regard to category.

Theorem 4. *The cardinality of the set of Gilbreath continued fractions, $\mathcal{G}$, is $\aleph_0$, the cardinality of the natural numbers.*

Proof. For any fixed N, there are 2^{N-1} possible Gilbreath permutations of the natural numbers up to N. Thus the set of numbers in $\mathcal{G}_F$ that look like

$$[0; \pi(1), \pi(2), \ldots, \pi(N)]$$

for each fixed natural number N is finite. Then the cardinality of $\mathcal{G}_F$ is $\aleph_0$, since it is a countable union of finite sets. Now, if $x \in \mathcal{G}_I$, there is a $k \in \mathbb{N}$ such that if we write $x = [0; a_1, a_2, \ldots, a_j, \ldots]$, then for $j \geq k$, we have $a_j = j$. In fact, this k is where $\pi(k-1) = 1$. So if we fix $N \in \mathbb{N}$ and insist $a_j = j$ for all $j > N$, how many prefixes are there for this continued fraction that are still Gilbreath? The answer is again 2^{N-1}. Thus there are finitely many sequences for each place where the continued fraction straightens out. So $\mathcal{G}_I$ is countable, too, which means $\mathcal{G}$ is a countable set. $\qquad\square$

An immediate consequence of this theorem is that the set $\mathcal{G}$ must be a first category set. That is, it is the countable union of nowhere dense sets. However, we can say more. The set $\mathcal{G}$ is, in fact, a *scattered* set, as defined by Freiling and Thomson [3].

Definition 3. Let $S \subset \mathbb{R}$. We say S is *scattered* if every nonempty subset of S contains an isolated point.

Scattered sets are distinct from countable sets and nowhere dense sets. The Cantor Set is nowhere dense (and uncountable) but not scattered. The rational numbers are countable but not scattered. Of course, a scattered set cannot be dense, but could be first category. However, it is known that scattered sets are both countable and nowhere dense [1].

Freiling and Thomson [3] prove that any countable G–delta set of real numbers is scattered.

Theorem 5. *The set of Gilbreath continued fractions, $\mathcal{G}$, is a scattered set in $\mathbb{R}$.*

Proof. For any $x = [0; a_1, a_2, a_3, \ldots, a_n] \in \mathcal{G}_F$ where n is fixed, we have that x must be isolated. For any $t \in \mathcal{G}_F$, let us assume t has length at most n. There are finitely many of these, so there exists an $\varepsilon_1 > 0$ such that the open ball (interval) centered at x with radius ε_1 does not intersect $\mathcal{G}_F$. If we were to append more numbers onto x, creating

$$y_k = [0; a_1, a_2, a_3, \ldots, a_n, n+1, n+2, \ldots, n+k],$$

then these y_k are converging toward some irrational number

$$y = [0; a_1, a_2, a_3, \ldots, a_n, n+1, n+2, \ldots, n+k, \ldots].$$

Thus there is an $\varepsilon_2 > 0$ such that the open ball centered at x with radius ε_2 intersects only finitely many y_k. This argument also explains why there is an ε_3 so that the open ball around x of that radius must miss $\mathcal{G}_I$. Thus the point is isolated.

If $x \in \mathcal{G}_I$, then x is not an isolated point in $\mathcal{G}$, as there is a sequence of points in $\mathcal{G}_F$ that converges to x. However, there is an $\varepsilon > 0$ so that

$$B(x, \varepsilon) \cap \mathcal{G}_I = \emptyset,$$

where $B(x, \varepsilon)$ is the open ball with center x and radius ε. This argument has to do with the place k where the continued fraction expansion straightens out; that is, for $j \geq k$ we have $a_j = j$. ☐

Topologically, it is clear that $\mathcal{G}$ is not open, since no scattered set in $\mathbb{R}$ is open. It is, in fact, a closed set in the unit interval.

Theorem 6. *The set of Gilbreath numbers is a closed set in $[0, 1]$.*

Proof. Pick an x in the complement of $\mathcal{G}$. This x has a continued fraction expansion that is *not* Gilbreath. Let N represent the least index where we see it is not Gilbreath; that is, $a_1, a_2, \ldots, a_{N-1}$ is ordered in $\{1, 2, 3, \ldots, N-1\}$. We create the following sequence of continued fractions:

$$y_N = [0; a_1, a_2, \ldots, a_{N-1}, b_N],$$
$$y_{N+1} = [0; a_1, a_2, \ldots, a_{N-1}, b_N, b_{N+1}],$$
$$y_{N+2} = [0; a_1, a_2, \ldots, a_{N-1}, b_N, b_{N+1}, b_{N+2}],$$
$$\vdots \qquad\qquad\qquad \vdots$$

where each y_n is a Gilbreath continued fraction.

Let

$$\varepsilon = \frac{1}{2}\left(\inf_{n \geq N}\{|x - y_n|\} \right).$$

Such an infimum must be positive, since there is a unique limit for the y_n (which is not x). Then the open ball with center x and radius ε is contained in the complement of $\mathcal{G}$. Thus $\mathcal{G}$ is a closed set. $\square$

We have derived some properties of the set of Gilbreath continued fractions. There is much more to consider, and many generalizations that could be constructed, but let us save that for another time.

References

[1] A. Bruckner, J. Bruckner and B. Thomson. *Real Analysis*. Prentice-Hall, 1997.

[2] P. Diaconis and R. Graham. *Magical Mathematics: The Mathematical Ideas that Animate Great Magic Tricks*. Princeton University Press, Princeton, NJ, 2011.

[3] C. Freiling and B. S. Thomson. Scattered sets and gauges. *Real Anal. Exchange* **21** no. 2 (1995–1996) 701–707.

[4] N. Gilbreath. Magnetic colors. *Linking Ring* **38** no. 5 (1958) 60.

[5] C. D. Olds. *Continued Fractions*. Mathematical Association of America, Washington, DC, 1963.

PART IV

◇◇◇◇◇◇◇◇

Games

13

◇◇◇

TIC-TAC-TOE ON AFFINE PLANES

Maureen T. Carroll and Steven T. Dougherty

Mathematicians often find nuggets of gold when mining recreational problems. The history of mathematics has numerous examples of puzzles, games, or unexpectedly interesting observations that blossomed into significant mathematical developments when placed under the microscope of a skilled problem-solver. Blaise Pascal (1623–1662) and Pierre de Fermat (1601–1665) famously developed the foundations for probability theory while corresponding about the resolution to a gambler's dispute. In 1852, a simply phrased question about coloring regions on a cartographer's map piqued the curiosity of Augustus De Morgan (1806–1871) and Arthur Cayley (1821–1895), inspiring more than a century's worth of mathematical results regarding the so-called "Four-Color problem". The greatest mathematician of all, Leonhard Euler (1707–1783), *started* the field of graph theory with the following navigational question: Is there a path through Königsburg that crosses every bridge exactly once [9]? Thankfully, this was not Euler's only foray into recreational mathematics, for it is with another of his investigations that we find the inspiration for our game.

In the last years of his life, Euler's examination of magic squares led him to so-called "Graeco-Latin squares." His findings in this area led to a paper of more than one hundred pages [10]. In this 1782 paper, Euler begins with the thirty-six-officer problem and ends with a conjecture about the possible sizes of Graeco-Latin squares. Despite the attempts of many mathematicians, this problem went unsolved for nearly two hundred years. In those two centuries, the mathematical ripple effect from his conjecture motivated generations of mathematicians to explore its connections to groups, finite fields, finite geometries, codes, and designs. In this chapter, we introduce a new game of tic-tac-toe that fits squarely within the body of work inspired by Euler's vision and the surprisingly interesting problem of arranging thirty-six officers of six different ranks and regiments. In an appropriate turn of events, this game brings us back full circle to the realm of recreational mathematics. First, we explain how to play the game, then we analyze it from a game-theoretic perspective to determine winning and drawing strategies. Along the way, we explain Euler's connection to the story.

1 Describing the Game

Our new game is built from two main ingredients. The first ingredient is familiar to all: the standard game of tic-tac-toe. It provides the rules for our new game. Two players take turns placing their mark in an unclaimed cell of a square grid. The first player to claim all cells in a straight line wins the game. If all cells are claimed and no player has won, then the game ends in a draw. The second ingredient puts us in Euler's realm and provides the game board: a finite affine plane.

An *affine plane* consists of a set of points and a set of lines. We require the points and lines to satisfy three axioms. If we let P denote the set of points (assumed to be nonempty), and if we let L denote the set of lines, then we assume the following.

- A-1: Through any two distinct points there exists a unique line.
- A-2: If $p \in P$, $\ell \in L$, and p is not contained in the line ℓ, then there exists a unique line m that passes through p and is parallel to ℓ. (Note: Distinct lines are parallel when they have no common points.)
- A-3: There are at least two points on each line, and there are at least two lines.

Notice that axioms A-1 and A-2 are Euclidean axioms. Axiom A-2 (the Parallel Postulate) guarantees that we are working in a plane, and axiom A-3 eliminates trivial geometries. Everyone will immediately notice that the Cartesian plane satisfies these axioms, and it is an example of an infinite affine plane. When we allow for a finite number of points, the axioms lead us to finite affine planes.

Suppose $p \in P$ and $q \in P$, and we have a line ℓ that consists of only these two points. We can use set notation $\{p, q\}$, geometric notation $\overline{pq}$, or a graph

p ●——————● q to represent line ℓ. When p is a point on line ℓ, we say that p is incident with ℓ. Note that we connect p and q with a segment to demonstrate that both points are incident with ℓ, but unlike in the Cartesian plane, there are no points in between p and q that are incident with ℓ, since this line contains only these two points.

Let's try to construct the smallest finite affine plane. Is this line ℓ with its two points an affine plane? No, axiom A-3 is not satisfied. If we add another line incident with p, and another point r on this line to satisfy axiom A-3, we have the graph in Figure 13.1. We ask ourselves again: Does this graph represent an affine plane? No, axiom A-1 is not satisfied since there is no line between r and q.

Figure 13.1. Is this an affine plane?

Figure 13.2. The smallest affine plane.

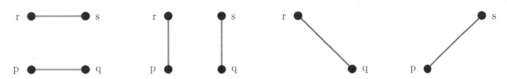

Figure 13.3. The six lines of the smallest finite affine plane.

If we continue in this fashion, trying to add as little as possible to satisfy the axioms, we will eventually construct the smallest affine plane. How many points do we need? How many lines would we have? A graph that represents the smallest finite affine plane we could construct is shown in Figure 13.2. The reader should verify the axioms for this plane.

There are four points and six lines, with each line containing exactly two points. The six lines found in the graph are shown individually in Figure 13.3. Notice that the three pairs, grouped by color, are parallel lines since they have no common points.

It's fun to try to construct these planes from the axioms. Starting with the smallest affine plane in Figure 13.2 and adding just one more point to one of the lines sets us on a path to constructing the next-largest finite plane. Try it! Be sure to keep track of the incidence structure and try to add the smallest possible number of points while adhering to the axioms. When successful, you will produce a graph of the second-smallest affine plane, isomorphic to the representation given in Figure 13.4. This plane has nine points and twelve lines, with three points on each line. The twelve lines found in the graph are shown individually in Figure 13.5. Notice that each of the four sets of lines grouped by color consists of mutually parallel lines.

Not that we are suggesting the constructive approach for the next-sized plane, but if we were to add another point and follow the axioms again, then Figure 13.6 shows a graph representing the next plane.

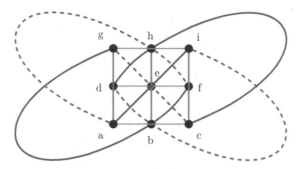

Figure 13.4. The second-smallest affine plane.

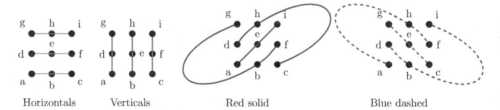

| Horizontals | Verticals | Red solid | Blue dashed |

Figure 13.5. The twelve lines of the second-smallest affine plane.

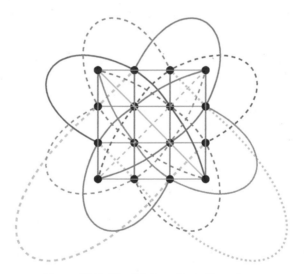

Figure 13.6. The third-smallest affine plane.

This plane has sixteen points and twenty lines, with four points on each line. The twenty lines found in the graph are shown individually in Figure 13.7. Take a moment to identify the five sets of mutually parallel lines.

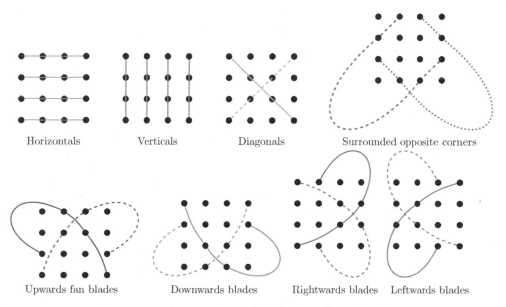

Figure 13.7. The twenty lines of the third-smallest affine plane.

Figure 13.8. The six ways to win in 2 × 2 tic-tac-toe.

If you've started thinking about how we will play tic-tac-toe on these planes, then you have surely noticed that these first three finite affine planes have a square number of points that can be arranged in the usual tic-tac-toe board shape. When we play on a finite affine plane, the points become the open cells in the $n \times n$ tic-tac-toe board, and we win if we can claim all points on a line in the affine plane. The geometry of the plane provides the twist on standard tic-tac-toe: lines here do not necessarily appear straight. Let's see how this works on the smallest plane we constructed, where there are two points on each line.

Consulting Figure 13.2 and Figure 13.3, we see the six winning lines on the 2 × 2 board are those given in Figure 13.8.

This game on the 2 × 2 board does not deliver on our promise of a new game, since it is the same as standard 2 × 2 tic-tac-toe. The first player, X, wins on his second move, as seen in Figure 13.9.

Our game gets a bit more interesting on the 3 × 3 board, where we have twelve winning lines instead of the standard eight. Consulting Figures 13.4 and 13.5, we see the twelve winning lines on the 3 × 3 board are those given

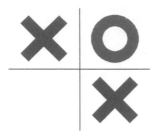

Figure 13.9. Tic-tac-toe on the 2 × 2 board.

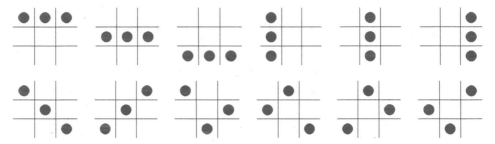

Figure 13.10. The twelve ways to win in 3 × 3 tic-tac-toe.

in Figure 13.10. The first eight lines are the usual horizontal, vertical, and diagonal winners. The last four lines are where we deliver on our promise of a new game. Notice that each of these four new winning lines requires exactly one mark in each row and column.

At this point, take a moment to play a few games to become accustomed to the new winning lines and to determine the best playing strategy. In Figure 13.11, we see a game in progress, where Player O must make a move. Here, even if Player O takes the lower left corner, Player X can win on the next move.

As we play and try to determine the best move for a given player at different stages of the game, we begin to notice properties of the cells (points) and winning arrangements (lines). For example, if Player X has two marks on the board, then these marks lie on exactly one winning line. This is the result of axiom A-1 for affine planes. We should take a moment to note some easily verified facts about finite affine planes, since these provide insights into playing the game. We can use the first three planes as our guide to lead us to general properties by answering such questions as: How many points are there? How many lines? How many points on each line? How many sets of parallel lines? Take a look at Figures 13.2 through 13.7 to answer these questions. The answers can be found in Figure 13.12 and the name and symbol for each plane is given at the bottom of each panel in the figure.

Figure 13.11. Player O's turn.

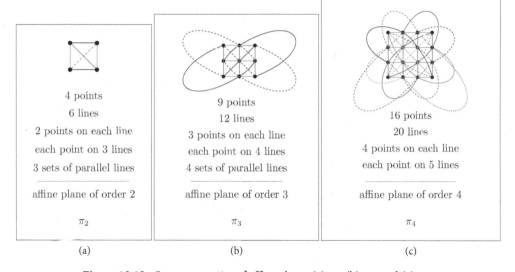

4 points

6 lines

2 points on each line

each point on 3 lines

3 sets of parallel lines

affine plane of order 2

π_2

(a)

9 points

12 lines

3 points on each line

each point on 4 lines

4 sets of parallel lines

affine plane of order 3

π_3

(b)

16 points

20 lines

4 points on each line

each point on 5 lines

affine plane of order 4

π_4

(c)

Figure 13.12. Some properties of affine planes (a) π_2, (b) π_3, and (c) π_4.

In general, an affine plane of order n has the following properties, which can be verified starting from the given axioms: there are n^2 points, $n^2 + n$ lines, n points on each line, $n + 1$ sets of mutually parallel lines with each set containing n lines, and each point in the plane is incident with $n + 1$ lines. Parallelism forms an equivalence relation on the set of lines, and each set of n mutually parallel lines is called a *parallel class*.

Let's play tic-tac-toe on π_4. Here we have a 4×4 board with twenty winning lines. Since it is easy to spot the four horizontal, four vertical, and two diagonal winners, we will just show the ten unusual winning lines. It will be helpful to

 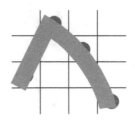

Figure 13.13. Winning line resembling a shark fin.

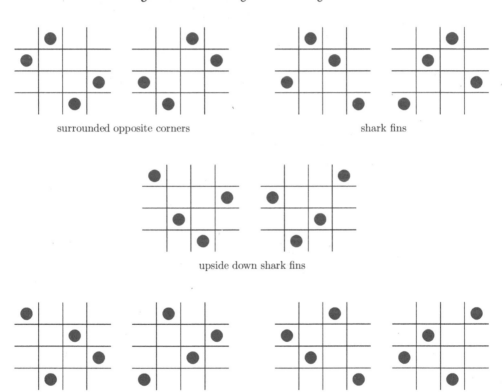

surrounded opposite corners   shark fins

upside down shark fins

rightwards pointing shark fins   leftwards pointing shark fins

Figure 13.14. New winning lines for 4×4 tic-tac-toe on π_4.

give visual descriptors for these winning arrangements based on their shape. As demonstrated in Figure 13.13, most of the new lines share a similarity with a shark fin.

Consulting Figures 13.6 and 13.7, we see the ten new winning lines on the 4×4 board as shown in Figure 13.14. If you use a pencil to sketch in the shark fins, you will find them pointing in all four directions.

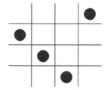

Figure 13.15. Not a winning line for 4×4 tic-tac-toe on π_4.

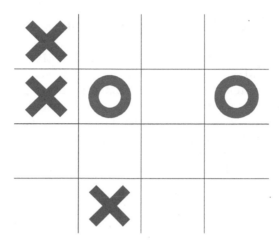

Figure 13.16. The first five moves of a sample 4×4 tic-tac-toe game on π_4. The ten unusual winning lines are shown in Figure 13.14.

Notice that, once again, each of the ten new winning lines requires exactly one mark in each row and column. However, this is not a sufficient condition for a winning line when playing on this plane. For example, in Figure 13.15, we see a configuration of marks meeting this condition but failing to be a winning line.

Figure 13.16 shows a 4×4 game board with the first five moves of a sample game given; the ten winning lines that are not immediately obvious are shown in Figure 13.14. After you have played enough games to identify wins and draws, it is natural to ask the following question. Can the first player always force a win in this game? This is not easy to answer. Let's delay the answer to this question and ask another instead.

Can we play this version of tic-tac-toe on any $n \times n$ sized board? Since there are values of n for which affine planes do not exist, the answer to this question is "no". For example, there is a π_5, but no affine plane of order 6 exists. Its nonexistence is specifically related to Euler's thirty-six officer problem, and, more generally, to orthogonal Latin squares.

$$\begin{bmatrix} 0 & 1 \\ 1 & 0 \end{bmatrix} \quad \begin{bmatrix} 0 & 1 & 2 \\ 2 & 0 & 1 \\ 1 & 2 & 0 \end{bmatrix} \quad \begin{bmatrix} 0 & 1 & 2 \\ 1 & 2 & 0 \\ 2 & 0 & 1 \end{bmatrix} \quad \begin{bmatrix} 0 & 1 & 2 & 3 \\ 1 & 0 & 3 & 2 \\ 2 & 3 & 0 & 1 \\ 3 & 2 & 1 & 0 \end{bmatrix} \quad \begin{bmatrix} 0 & 1 & 2 & 3 \\ 2 & 3 & 0 & 1 \\ 3 & 2 & 1 & 0 \\ 1 & 0 & 3 & 2 \end{bmatrix} \quad \begin{bmatrix} 0 & 1 & 2 & 3 \\ 3 & 2 & 1 & 0 \\ 1 & 0 & 3 & 2 \\ 2 & 3 & 0 & 1 \end{bmatrix}$$

Figure 13.17. Some Latin squares of orders 2, 3, and 4.

2 Affine Planes and Latin Squares

Consider the thirty-six officer problem: There are thirty-six officers, each with one of six ranks and one of six regiments. Can these officers be arranged into six rows and six columns so that each rank and regiment is represented in each row and each column? Leonhard Euler showed this could be done for nine, sixteen, and twenty-five officers but conjectured correctly that it could not be done for thirty-six. In an attempt to solve this problem, he introduced Latin squares [10]. A *Latin square* of order n is an $n \times n$ matrix with entries from $\mathbb{Z}_n = \{0, 1, 2, \ldots, n-1\}$, where each number occurs exactly once in each row and each column.

Latin squares $A = [a_{ij}]$ and $B = [b_{ij}]$ are said to be *orthogonal* if $C = [c_{ij}]$, whose entries are the ordered pairs $c_{ij} = (a_{ij}, b_{ij})$, contains all n^2 possible ordered pairs of $\mathbb{Z}_n \times \mathbb{Z}_n$. A collection of Latin squares is said to be *mutually orthogonal* if and only if each pair is orthogonal. We use the abbreviation "MOLS" for "Mutually Orthogonal Latin Squares." In Figure 13.17, the two Latin squares of order 3 are orthogonal, and the three of order 4 are MOLS. Euler's thirty-six officer problem asks whether it is possible to find a pair of orthogonal Latin squares of order 6, one representing the ranks of the thirty-six officers and the other representing the regiments. As illustrated by the first Latin square of order 3 in Figure 13.17, you can easily produce one Latin square of order 6 by continually shifting the elements of your first row to the right by one position and wrapping the leftover elements to the beginning. The proof of the thirty-six officer problem shows that you cannot produce a second Latin square orthogonal to the first. Exhaustive solutions to this problem [16], as well as more sophisticated proofs [7, 15], can be found in the literature.

Given an order n, it is known that there can be at most $n - 1$ MOLS. When this upper bound is achieved, we have a complete set of orthogonal Latin squares of order n. R. C. Bose (1901–1987) showed that an affine plane of order n exists if and only if there is a complete set of orthogonal $n \times n$ Latin squares [5]. So the existence of an affine plane of order 6 relies on the existence of five MOLS of order 6, but as we have noted, there aren't even a pair of these orthogonal squares! Thus, it is impossible to play our new tic-tac-toe game on a 6×6 board.

What sized boards *are* possible?

It is well known that there are affine planes of order p^k, where p is prime and $k \in \mathbb{Z}^+$. This tells us, for example, that there are affine planes of orders 2, 3, 4, 5, 7, 8, 9, and 11. The proof of the nonexistence of the plane of order 10 requires a great deal of mathematics and an enormous computation to finish the proof. (See Lam [12] for a historical account.) The existence of an affine plane of order 12 is an open problem. In fact, it is unknown whether there are any affine planes that do not have prime-power order. Some of these orders, however, are known not to exist (see the Bruck-Ryser Theorem [2]). For our game, the important thing to note here is that there are infinitely many planes on which we can play tic-tac-toe. However, producing a graph to represent the winning lines on planes of higher order is impractical. As we describe below, the MOLS will work nicely here as a substitute for a graph.

When we venture beyond the planes of small order, the connection between affine planes and complete sets of orthogonal Latin squares can be used to identify lines of the plane quite easily. After arranging the n^2 points of a finite affine plane in an $n \times n$ grid, we first identify its $n + 1$ parallel classes, which in turn reveals all of the lines. The n horizontal lines form one parallel class, and the n vertical lines form another. Each of the remaining $n - 1$ parallel classes corresponds to one of the $n - 1$ MOLS as follows: the ith line in any parallel class is formed by the positions of symbol i in the corresponding Latin square. (Here, $i = 0, 1, \ldots, n - 1$.) With a familiar small plane as an example, using Figure 13.4 and the two orthogonal 3×3 Latin squares in Figure 13.17, we see that the four parallel classes for the affine plane of order 3 are

(i)   the horizontal lines $\{\{a, b, c\}, \{d, e, f\}, \{g, h, i\}\}$,
(ii) the vertical lines $\{\{a, d, g\}, \{b, e, h\}, \{c, f, i\}\}$,
(iii) the lines indicated by identical numbers in the first Latin square

$$\{\{c, e, g\}, \{a, h, f\}, \{i, b, d\}\}, \text{ and}$$

(iv) the lines indicated by identical numbers in the second Latin square

$$\{\{g, b, f\}, \{c, h, d\}, \{a, e, i\}\}.$$

Take a moment to label the points in Figure 13.6, and then use the complete set of orthogonal 4×4 Latin squares given in Figure 13.17 to identify the lines of the five parallel classes of π_1. Using the MOLS to identify the unusual (nonhorizontal or nonvertical) winning lines on planes of higher order is the only reasonable way to play tic-tac-toe, since the graphs for these planes get a bit unwieldy. To produce a complete set of Latin squares, the Maple® command *MOLS(p,m,n)* provides n MOLS of order p^m when p is prime and $n < p^m$. Now that we have a way to identify the winning lines of the infinitely many planes on which we can play our game of tic-tac-toe, let's analyze the game to determine the best way to play.

3 Winning and Drawing Strategies

A *zero-sum game* is a game in which one player's loss is a gain for the other player(s). Tic-tac-toe on π_n is a two-player, zero-sum game on an $n \times n$ board where the cells in the grid are identified with the points on π_n. Players alternately mark one open cell with an X or an O. For simplicity, assume that Player X makes the first move. A player wins by being the first to claim n cells on a line in π_n. If a game is complete and no player has won, the game is a draw. Tic-tac-toe is an example of a game of perfect information, since each choice made by each player is known by the other player. Poker is not such a game, since players do not reveal their cards.

A *strategy* is an algorithm that directs the next move for a player based on the current state of the board. A winning strategy for Player X, for example, is a strategy that is guaranteed to produce a win for him. For example, consider a variation on Nim known as the Matchsticks or Subtraction game, where two players take turns removing up to three matches from an initial pile of ten matches. The first player's winning strategy is to leave her opponent with 1 modulo 4 matches at every turn. As an alternate example, while there is a best way to play the standard 3×3 tic-tac-toe game we learned as children, here no winning strategy exists, since each player can guarantee that the other cannot win. In this case, we say that both players have a *drawing strategy*, that is, an algorithm that leads to a draw. The assumption that both players are knowledgeable and play correctly is a standard game-theoretic assumption called the *principle of rationality*, that is, at each move, each player will make a choice leading to a state with the greatest utility for that player. To indicate the order of play, we denote Player X's first move as X_1, her second move as X_2, and so on. Player O's moves are analogously designated.

In game theory it is known that in a finite two-player game of perfect information, either one player has a winning strategy or both players can force a draw [14]. A strategy-stealing argument [3, 4] proved by Hales and Jewett [11] shows that in our case, it is the first player who has a winning strategy when such a strategy exists. To show this, assume that Player O has the winning strategy. Let Player X make a random first move and thereafter follow the winning strategy of Player O. Specifically, Player X plays as if she *were* Player O by pretending that her first move has not been made. If at any stage of the game she has already made the required move, then a random move can be made. Any necessary random moves, including the first, cannot harm her, since she is merely claiming another open cell in the grid. This leads Player X to a win, contradicting the assumption that Player O has the winning strategy. (Notice that this argument does not apply to Nim, e.g., since a random move may cause the first player to lose.) Hence, in tic-tac-toe either the first player has a winning strategy or both players have drawing strategies, in which case we say the second player can force a draw. If no draws exist, then the first player is

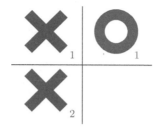

Figure 13.18. Game played on π_2.

guaranteed to have a winning strategy. However, the existence of draws is not enough to guarantee that the second player can force a draw.

We now discuss the existence of a winning strategy for Player X or a drawing strategy for Player O in all finite affine planes in the two sections that follow.

4 The 2×2 and 3×3 Boards

As we have seen earlier, on a 2×2 board, even Euler would lose to a randomly pecking chicken if he let the chicken make the first move. As soon as the chicken claims X_2, the game is over, since the line joining X_1 and X_2 is complete. A representative example is shown in Figure 13.18. So, the first player has a winning strategy and no draws exist on π_2.

There is a unique affine plane of order 3 as represented in Figure 13.4. Player X has a winning strategy on this plane as follows. Since every point is incident with the same number of lines, and since there is a line between any two points, we may assume that X_1 and O_1 are chosen arbitrarily. Player X then chooses X_2 to be any point not on the line containing X_1 and O_1. In the representative game shown in Figure 13.19a, Player X may choose any open cell except the one in the top row.

Since there is a line between any two points, O_2 must block the line containing X_1 and X_2 to prevent Player X from winning on her next move. Likewise, Player X must choose X_3 to be the point on the line containing O_1 and O_2. Our representative example has now progressed to the state shown in Figure 13.19b. At this stage of game play, Player O must block either the line containing X_1 and X_3 or the line containing X_2 and X_3, (it is a simple matter to see that he has not already blocked these lines). Player X wins on her next move when she completes the line that O_3 did not block.

If the principle of rationality is violated then the game could end in a win for the second player, but a draw is impossible, since there are no draws on the 3×3 board. To show this, assume that a draw *is* possible and let D be the

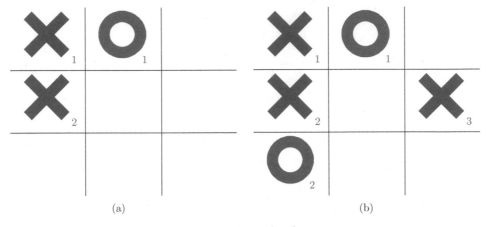

(a)   (b)

Figure 13.19. Game played on π_3.

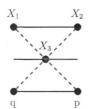

Figure 13.20. There are no draws on π_3.

set of five points that meets each line in the plane without containing any line completely. Let ℓ_1, ℓ_2, and ℓ_3 be the three lines of one of the parallel classes of π_3. Without loss of generality, assume D meets both ℓ_1 and ℓ_3 at two points and ℓ_2 at one point. Let D meet ℓ_1 at points X_1 and X_2, and ℓ_2 at point X_3. The line between X_1 and X_3 intersects ℓ_3, say, at p. The line between X_2 and X_3 also intersects ℓ_3, say, at q. Figure 13.20 illustrates these facts.

The two dashed lines are incident with X_3, and ℓ_3 is not parallel to either of these lines. Therefore, we have $p \neq q$, since no two lines intersect in more than one point. Since ℓ_3 has exactly three points and D intersects ℓ_3 in two points, if p is not in D, then q must be in D. So, either line $\{X_1, X_3, p\}$ or $\{X_2, X_3, q\}$ is in D, which contradicts our assumption of the existence of a draw.

To recap our findings on the 2×2 and 3×3 boards: the first player has a winning strategy, and no draws exist on π_2 and π_3.

5 Weight Functions and Larger Boards

When we venture beyond the planes of small order, the complexity of the game increases dramatically. The additional points and lines generate a far greater number of possible moves for each player. This prevents an easy move-by-move analysis as we carried out in the previous section. This is where Paul Erdős (1913–1996) comes to our rescue. The two theorems that follow are special cases of a result of Erdős and Selfridge [8] specifying conditions under which the second player can force a draw in many positional games. Our proofs are a modification of the proof of the Erdős and Selfridge theorem given by Lu [13].

To analyze the game on any plane of order n, we need a way to evaluate the state of the board at a given stage of play. It would be helpful to assign a number that in some way measures the utility of the state of the board for one of the players. To do this, we define functions that assign values to various elements of the board when Player O is about to make his ith move. To choose the position for O_i from the unclaimed points remaining, he may first wish to consider which line has the best available point. As the second player, the best outcome he can achieve under the assumption of rational play is a draw. Also, he forces a draw if he places one of his marks on every line in the plane, thereby blocking every possible winning line for Player X. So, any line that already has at least one O is blocked and should contribute no weight. Of the unblocked lines remaining, it is most important for him to block one of the lines with the largest number of Player X's marks. If we define the value, or weight, of an unblocked line to be 2^{-a}, where a is the number of available, or unmarked, points on that line, then the lines of greater weight are precisely the lines that have more Xs, and are, therefore, urgent for him to block. As Player O is about to make his ith move, the *weight of the board* is defined as the sum of the weights of the lines in the plane. The *weight of an available point* is the sum of the weights of the lines incident with the point. Lastly, the *weight of a pair of available points* is the weight of the line through these points.

The state of the board changes with game play. We use B_i to refer to the state of the board just before O makes his ith move. Since the index of the final board state depends on both the order of the plane and progress of play, we let B_∞ denote the board when no more moves can be made. That is, the game has ended in a win or a draw.

If p and q are available points and ℓ is a winning line in the plane, then we have

$$\text{Weight of line } \ell \text{ blocked by O} = w(\ell) = 0,$$

$$\text{Weight of line } \ell \text{ unblocked by O} = w(\ell) = 2^{|\text{available points on } \ell|},$$

$$\text{Weight of the board immediately before move } O_i = w(B_i) = \sum_\ell w(\ell),$$

$$\text{Weight of available point } q \text{ at board state } B_i = w(q|B_i) = \sum_{\ell \text{ through } q} w(\ell),$$

and

$$\text{Weight of available pair } \{p, q\} \text{ at board state } B_i = w(p, q|B_i) = w(\overline{pq}),$$

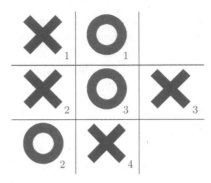

Figure 13.21. B_∞ for a game played on π_3.

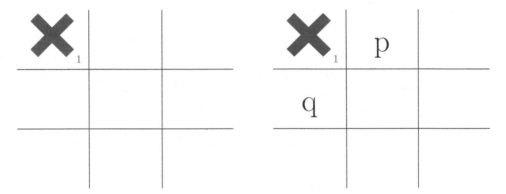

Figure 13.22. B_1 for a game played on π_3.

where $\overline{pq}$ is the unique line through p and q. Let's look at some calculations using the final board state B_∞ of a game played on π_3, as given in Figure 13.21.

As shown in Figure 13.22, board state B_1 is directly before move O_1, meaning only X_1 has been played. At this stage, all twelve lines are unblocked, the four lines through X_1 have two available points, and the eight remaining lines have three available points. This gives $w(B_1) = 4 \cdot 2^{-2} + 8 \cdot 2^{-3}$.

Since there are four lines through any point on π_3 and there is a unique line between X_1 and any other point, at this stage of game play, all available points have equal weight. In particular, through either p or q, there are three unblocked lines with three available points, and one line through X_1 with two available points. This gives $w(p|B_1) = w(q|B_1) = 2^{-2} + 3 \cdot 2^{-3}$. Since point p becomes O_1 and point q becomes X_2 later in this game, we use $w(O_1|B_1)$ and $w(X_2|B_1)$ to refer to the weights of these particular points at board state B_1, even though they have not been played yet. So, here we have $w(O_1|B_1) = w(X_2|B_1) = 2^{-2} + 3 \cdot 2^{-3}$. Also, we have $w(X_2, O_1|B_1) = w(p, q|B_1) = 2^{-3}$,

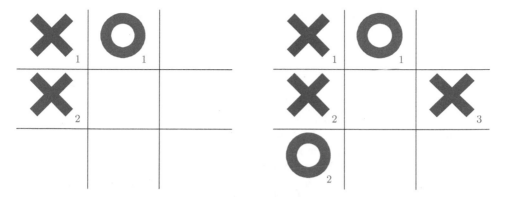

Figure 13.23. B_2 and B_3 for a game played on π_3.

since the line through the points represented by X_2 and O_1 has three available points at board state B_1. Board states B_2 and B_3 are shown in Figure 13.23.

To calculate $w(B_2)$, we may ignore the four lines blocked by O_1, since they contribute no weight. The eight remaining unblocked—and therefore weight-contributing—lines are as follows.

Three lines through X_1: one line through X_2 has one available
 point, and the other two have two available points $2^{-1} + 2 \cdot 2^{-2}$
Two uncounted lines through X_2 (the third line through X_1
 has already been counted), each with two available points $2 \cdot 2^{-2}$
Three remaining uncounted lines, each with three available
 points $3 \cdot 2^{-3}$

Thus we have $w(B_2) = 2^{-1} + 4 \cdot 2^{-2} + 3 \cdot 2^{-3}$. At this stage of game play, available points vary in weight. For example, at B_2, the weights of the points later marked by O_2 and X_3 are $w(O_2|B_2) = 2^{-1} + 2 \cdot 2^{-3}$ and $w(X_3|B_2) = 2 \cdot 2^{-2} + 2^{-3}$. (The reader can verify that the weight of any available point on board B_2 will be $\frac{1}{2}$, $\frac{5}{8}$, or $\frac{3}{4}$.) The weight of this particular pair of available points at board state B_2 is $w(X_3, O_2|B_2) = 0$, since it is blocked by O_1. Finally, when considering B_3, we eliminate the seven lines blocked by O_1 or O_2 from consideration, leaving $w(B_3) = 2 \cdot 2^{-1} + 3 \cdot 2^{-2}$.

Consider the difference in weights between two successive states of the board, $w(B_i) - w(B_{i+1})$. The only changes between board states B_i and B_{i+1} are the addition of Player O's ith move and X's $(i+1)$st move. So, the weights of any lines that do not contain O_i and X_{i+1} do not change and therefore, cancel each other out. With only the lines through either of these two points remaining, the weights of the lines through X_{i+1} must be subtracted from the weights of the lines through O_i to find $w(B_i) - w(B_{i+1})$. Since this eliminates the weight of the line that passes through both points, the weight of this line must be added back. Thus, it can be seen that

$$w(B_i) - w(B_{i+1}) = w(O_i|B_i) - w(X_{i+1}|B_i) + w(X_{i+1}, O_i|B_i). \quad (1)$$

The examples given above can be used to demonstrate equation (1) when $i = 1$ and $i = 2$.

These weight functions enable us to check for draws at any stage of the game. First notice that if Player X has completed a winning line, then the winning line is unblocked by Player O and the line has no available points. Thus, the weight contribution of a winning line to $w(B_\infty)$ is $2^{-0} = 1$. So as long as $w(B_i) < 1$, then Player X has not completed a line. Also, if $w(B_\infty) < 1$, then play has ended in a draw. Moreover, these weight functions provide strategies for both players and help us determine the outcome of play on all planes of higher order. Specifically, Player X should minimize $w(B_i) - w(B_{i+1})$ to keep the weight of B_j, at any stage j of the game, above 1, whereas Player O should maximize this difference to drag the overall weight below 1. Hence, by equation (1), the second player chooses O_i by maximizing $w(O_i|B_i)$, and the first player chooses X_{i+1} by maximizing $w(X_{i+1}|B_i) - w(X_{i+1}, O_i|B_i)$. The power and utility of these weight functions is demonstrated in the proof of the following theorem, where the drawing strategy for Player O is specified for infinitely many affine planes.

Theorem 1 (Draw Theorem for π_n). *The second player can force a draw on every affine plane of order n with $n \geq 7$.*

Proof. To prove that Player O can force a draw, we must produce an algorithm that prescribes his move at any stage of the game and then show that this strategy leads to a draw. As noted above, if $w(B_\infty) < 1$, then Player O has forced a draw. This is equivalent to showing that there exists N, with $1 \leq N < \infty$, such that $w(B_N) < 1$, and

$$w(B_{i+1}) \leq w(B_i) \text{ for all } i \geq N.$$

Suppose that the current state of the board is B_i and Player O is about to make his ith move. Since the weight function assigns more weight to lines on which Player X is closer to winning, Player O should choose a point of maximal weight to ensure that he chooses a point incident with lines that most urgently require blocking. So, choose O_i such that

$$w(O_i|B_i) = \max\{w(q|B_i) : q \text{ is available on board } B_i\}.$$

By the choice of O_i and equation (1), we see that the second condition is always satisfied, since $w(O_i|B_i) \geq w(X_{i+1}|B_i)$.

An affine plane of order n has n points on each line, $n^2 + n$ lines, and $n + 1$ lines through each point. So, B_1 consists of $n + 1$ lines through X_1 and

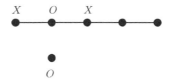

Figure 13.24. Drawing strategy on π_5: configuration of points after first four moves.

$(n^2 + n) - (n + 1)$ completely available lines. This gives

$$w(B_1) = (n + 1) \cdot 2^{(n-1)} + (n^2 - 1) \cdot 2^{-n} = \frac{n^2 + 2n + 1}{2^n}.$$

We see that $w(B_1) < 1$ when $n \geq 6$. Since the second condition guarantees that the weight of the board is nonincreasing, once the board weight dips below 1, we are assured that Player X cannot complete a winning line. Thus, by choosing a point of maximum weight at every stage of the game, Player O forces a draw on the affine planes of order $n \geq 7$ (since there is no such plane of order 6). $\square$

Since we have already shown that Player X has a winning strategy on π_2 and π_3, the only planes remaining for analysis are π_4 and π_5. It is interesting to note that we found greater difficulty determining the outcome of play on π_4 and π_5 than on planes of higher order. While we had hoped to determine the outcome of play on these planes without performing calculations for all possible outcomes by hand, the unsuspected complexity of play on π_4 and π_5 lent itself to a move-by-move analysis and machine computations.

Let's start with our findings on the 5×5 board. This board has thirty lines with five points on each line and six lines through each point. We calculate $w(B_3)$ after providing the strategy for Player O's first two moves. As previously noted, all boards with only two marks, X_1 and O_1, are isomorphic. Next, we assume Player X places X_2 anywhere. If O_1 is already on the line containing X_1 and X_2, then O_2 should not be placed on this line. If O_1 is not on the line containing X_1 and X_2, then O_2 should be placed on this line. In either case, the configuration before move X_3 is represented in Figure 13.24. The three collinear marks can have any arrangement on the indicated line. We only need to establish the incidence relationships to calculate weights.

The placement of X_3 leaves only four possible configurations of points, as represented in Figure 13.25. Without loss of generality, we have labeled the other points to aid in the calculations that follow. As long as $w(B_3) < 1$ in each case, then Player O has forced a draw.

In all four cases, we start by eliminating the six lines through O_1 and the remaining five lines through O_2, since these blocked lines contribute no weight. Once these eleven lines are eliminated from consideration, there are

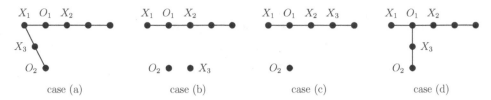

Figure 13.25. Drawing strategy on π_5: Four possible configurations for B_3.

nineteen unblocked lines remaining to be included in the weight function. We present the cases from smallest to largest weight.

Case (c): Of the nineteen unblocked lines, there are:

Four lines through X_1, each with four available points $\qquad$ $4 \cdot 2^{-4}$

Four uncounted lines through X_2, each with four available points $\qquad$ $4 \cdot 2^{-4}$

Four uncounted lines through X_3, each with four available points $\qquad$ $4 \cdot 2^{-4}$

Seven remaining, each with five available points $\qquad$ $7 \cdot 2^{-5}$

Thus we have $w(B_3) = 12 \cdot 2^{-4} + 7 \cdot 2^{-5} = \frac{31}{32}$.

Case (a): Of the nineteen unblocked lines, there are:

Four lines through X_1, each with four available points $\qquad$ $4 \cdot 2^{-4}$

Four uncounted lines through X_2: one line through X_3 has three available points, and the other three have four available points $\qquad$ $2^{-3} + 3 \cdot 2^{-4}$

Three uncounted lines through X_3 (the fourth line through X_2 has already been counted), each with four available points $\qquad$ $3 \cdot 2^{-4}$

Eight remaining, each with five available points $\qquad$ $8 \cdot 2^{-5}$

Thus we have $w(B_3) = 2^{-3} + 10 \cdot 2^{-4} + 8 \cdot 2^{-5} = 1$.

Case (b): The nineteen unblocked lines in this case give $w(B_3) = 2 \cdot 2^{-3} + 8 \cdot 2^{-4} + 9 \cdot 2^{-5} = \frac{33}{32}$.

Case (d): The nineteen unblocked lines in this case give $w(B_3) = 2 \cdot 2^{-3} + 9 \cdot 2^{-4} + 8 \cdot 2^{-5} = \frac{34}{32}$.

Player O forces a draw in case (a), but the others must be broken down into subcases to show that the weight of the board will fall below 1. We used a computer to verify that the second player can force a draw in the three remaining cases for π_5. The following theorem summarizes these results.

Figure 13.26. A draw on the affine plane of order 4.

Theorem 2 (Draw Theorem for π_n). *The second player can force a draw on every affine plane of order n with $n \geq 5$.*

Only one plane is left to consider. What happens on the affine plane of order 4? The game in Figure 13.26 shows that draws exist on π_4.

Since we had no examples of a plane for which a winning strategy and draws coexisted, it was natural to expect that the second player could force a draw. In the analysis of the same starting cases we used for π_5, the weights at board state B_3 on π_4 for cases (a) through (d) are $\frac{19}{16}$, $\frac{22}{16}$, $\frac{20}{16}$, and $\frac{23}{16}$, respectively. These numbers necessitated a further breakdown of each case, and the analysis of the subcases did not produce the expected result. To our surprise, three independent computer algorithms and an unpublished manuscript show that the first player has a winning strategy on this plane. The first two programs, written by students J. Yazinski and A. Insogna (University of Scranton, Pennsylvania), use a tree-searching algorithm. The third program, written by I. Wanless (Oxford University), checks all possible games up to isomorphism. The manuscript written by M. Conlen (University of Michigan, Ann Arbor) and J. Milcak (University of Toronto) gives a move-by-move analysis with many subcases. Thus, the affine plane of order 4 is the only plane for which the first player has a winning strategy, and yet, draws exist.

6 Further Exploration

Since projective planes are a natural extension of affine planes, we can also play tic-tac-toe on these planes. Our analysis of winning and drawing strategies on these planes has been published elsewhere [6]. In the event that you are the *second* player on a plane where a forced draw is possible, in addition to the computational drawing strategy demonstrated in this chapter, we also provide blocking configurations for these planes, that is, simple configurations of points that produce a draw with very few points.

At the University of Scranton we have held a single-elimination "Tic-tac-toe on π_4" tournament where students compete for prizes. We encourage the reader to play, too, as we have found that these students gain not only an understanding of affine planes but also develop an intuition for finite geometries that reveals properties and symmetries not easily seen by reading definitions in a geometry text. For further reading, surveys of other tic-tac-toe games can be found in Berlekamp, Conway, and Guy [3] and Beck [1].

References

[1] J. Beck. Achievement games and the probabilistic method, in *Combinatorics, Paul Erdős Is Eighty*, pp. 51–78. Vol. 1, Bolyai Society Mathematical Studies, János Bolyai Mathematics Society, Budapest, 1993.

[2] M. K. Bennett. *Affine and Projective Geometry*. Wiley, New York, 1995.

[3] E. R. Berlekamp, J. H. Conway, and R. K. Guy. *Winning Ways for your Mathematical Plays*, Volume 2. Academic Press, London and New York, 1982.

[4] K. Binmore. *Fun and Games: A Text on Game Theory*. D. C. Heath and Co., Lexington, MA, 1992.

[5] R. C. Bose. On the application of the properties of Galois fields to the problem of construction of hyper-Graeco-Latin squares. *Sankhya* **3** (1938) 323–338.

[6] M. T. Carroll and S. T. Dougherty. Tic-tac-toe on a finite plane. *Math. Mag.* **77** no. 4 (2004) 260–274.

[7] S. T. Dougherty. A coding theoretic solution to the 36 officer problem. *Des. Codes Cryptogr.* **4** (1994) 123–128.

[8] P. Erdős and J. L. Selfridge. On a combinatorial game. *J. Combin. Theory* **14** (1973) 298–301.

[9] L. Euler. Solutio problematis ad geometriam situs pertinentis. *Commentarii academiae scientiarum Petropolitanae* **8** (1741) 128–140. Reprinted in L. Euler, *Opera Omnia*, pp. 1–10, ser. 1, vol. 7. Tuebner, Berlin and Leipzig, 1923.

[10] L. Euler. Recherches sur une nouvelle espace de quarees magiques. *Verh. Zeeuwsch Genootsch. Wetensch. Vlissengen* **9** (1782) 85–239. Reprinted in L. Euler, *Opera Omnia*, pp. 291–392, ser. 1, vol. 7, Tuebner, Berlin and Leipzig, 1923.

[11] A. W. Hales and R. I. Jewett. Regularity and positional games. *Trans. Amer. Math. Soc.* **106** (1963) 222–229.

[12] C. Lam. The search for a finite projective plane of order 10. *Am. Math. Monthly* **98** (1991) 305–318.

[13] X. Lu. A characterization on *n*-critical economical generalized tic-tac-toe games. *Discrete Math.* **110** (1992) 197–203.

[14] A. Rapoport. *Two-Person Game Theory; The Essential Ideas.* University of Michigan Press, Ann Arbor, 1966.

[15] D. R. Stinson. A short proof of the nonexistence of a pair of orthogonal Latin squares of order six. *J. Combin. Theory A* **36** (1984) 373–376.

[16] G. Tarry. Le problème des 36 officiers. *Compt. Rend. Assoc. Franc. Avacem. Sci.* **2** (1901) 170–203.

14

◇◇

ERROR DETECTION AND CORRECTION USING SET®

Gary Gordon and Elizabeth McMahon

The card game SET® was introduced in 1991 and was immediately embraced by the game-playing public, winning multiple awards [5]. Mathematicians quickly realized the game provides a concrete model for four-dimensional affine geometry over the finite field $GF(3)$. The game allows teachers to give students a gentle and fun introduction to finite geometry, some elementary counting, probability, linear algebra, and a variety of other topics. Concentrating on connections to linear algebra leads to error detection and correction, culminating with a perfect linear code composed of the SET® cards. The game also serves as an attractive entry to several interesting research questions in finite geometry [2, 3].

SET® is played with a special deck of eighty one cards. Each card in the deck has symbols with four distinct attributes:

- Number: One, two, or three symbols;
- Color: Red, purple, or green;
- Shading: Empty; striped; or solid; and
- Shape: Ovals; diamonds; or squiggles.

A set consists of three cards for which each attribute is independently either all the same or all different. Figure 14.1 shows two sets and one non-set (i.e., a collection of three cards that is not a set, because two of the cards are ovals but the third is not).

In the play of the game, twelve cards are placed face up, and the first person who finds three cards that form a set calls "Set." That player removes those three cards, three new cards are placed face up, and the game continues.

If there are no sets among the twelve cards, three more cards are placed, bringing the total to fifteen. While it is possible that no sets will be present

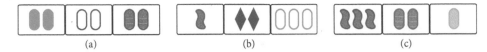

Figure 14.1. (a, b) Two sets; (c) one non-set.

among the fifteen cards, this is a fairly rare event. (Although we do not explore this question here, the maximum number of cards with no set is twenty.) The game continues until no sets can be taken. Typically, six or nine cards remain at the end of the game—computer simulations show that one of these two possibilities occurs about 90% of the time. The winner is the person who took the most sets.

Over the course of a game, a player may make a mistake, of course, by taking three cards that do not form a set. This event, which is fairly common among players learning the game, motivates the two fundamental questions we consider here.

- How can we detect whether an error has been made in playing the game?
- If an error has been made, can we correct it?

To answer these questions, we introduce coordinates for the cards and then use these coordinates to define a Hamming weight (see Definition 1 in Section 2) for any subset of cards. We will see that sets correspond precisely to subsets of size three with Hamming weight 0. But we will also see that the Hamming weight of the entire deck is 0, a very easy fact that has some important consequences.

In Section 1, we use these facts about Hamming weight to describe a variant of the game, called the *EndGame*, where one card is hidden at the beginning of play. Using the fact that subsets of weight 0 are removed (these are the sets) during the play of the game, we can determine the hidden card uniquely from the cards remaining on the board at the end of the game. If that hidden card forms a set with two of the cards remaining on the board, a player can call "Set," choose those two cards on the board, and then, dramatically, reveal the hidden card that makes a set with those cards.

The EndGame leads naturally to error detection in the game. When the card determined by the cards showing at the end of the game does not match the hidden card, an error must have been made in the course of play. In Section 2, we indicate how this procedure can be used to detect a single error. We also give some limitations on this "parity-check" method.

Classical linear codes are subspaces of finite-dimensional vector spaces over finite fields. Our coordinate system naturally associates the collection of eighty-one SET® cards with the vector space $\mathbb{F}_3^4$ (i.e., the collection of all ordered 4-tuples over $GF(3)$). An elementary argument then allows us to find a *perfect code* among these cards. We examine this connection to coding theory in

TABLE 14.1.
Assignment of coordinates to cards

i	Attribute	Value		a_i
1	Number	3,1,2	↔	0, 1, 2
2	Color	green, purple, red	↔	0, 1, 2
3	Shading	empty, striped, solid	↔	0, 1, 2
4	Shape	diamond, oval, squiggle	↔	0, 1, 2

Section 3, producing a perfect, single-error-correcting linear code solely from SET® cards. The cards in the code form a plane in the affine geometry, so we call them *code planes*, and we examine a few of their rather special properties.

We conclude in Section 4 with additional topics that demonstrate the deep connections between the simple card game and advanced mathematics. Every examination of the structural properties leads to new connections, discoveries, and incidence relations.

1 The EndGame

To use ideas from linear algebra, we need coordinates for our SET® cards. A card will have coordinates (a_1, a_2, a_3, a_4) determined as shown in Table 14.1. For instance, the card with two purple empty ovals would be represented by the 4-tuple $(2, 1, 0, 1)$. This assignment is obviously arbitrary, but, having made these choices, we fix them for the remainder of this chapter.

Once we have settled on our coordinates, we can treat the cards as vectors, so we can add two cards coordinatewise, modulo 3. With this in mind, we can define the Hamming weight of a subset of cards.

Definition 1. The *Hamming weight* $w(C)$ of a card C is the number of nonzero coordinates of the card. The weight $w(S)$ of a subset S of cards is the weight of the modulo 3 sum of the cards in S.

While the weight of a card depends on the arbitrary assignment of attributes to coordinates, the next result is independent of this assignment.

Proposition 1. Let S be a subset of three cards. Then $w(S) = 0$ if and only if S is a set.

It's easy to see that a set must have weight 0: either the attributes are all the same, so this coordinate contributes $3a_i \equiv 0 \pmod 3$ to the weight, or the attributes are all different, and so this coordinate contributes $0 + 1 + 2 \equiv 0 \pmod 3$ to the weight. Conversely, it is easy to check that a non-set of three cards has non-zero weight—we leave the details to the interested reader.

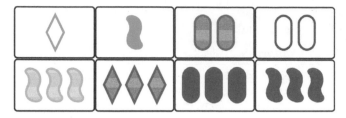

Figure 14.2. Cards left on the table at the end of the game. What's the missing card?

In addition, the entire deck has Hamming weight 0, since each expression for any attribute can be found on twenty-seven cards. One consequence of this fact is the following:

> Assuming no mistakes have been made in the play of the game, it is impossible for the game to end with exactly three cards left on the table.

Why is this true? We are removing sets of weight 0 from a deck that has weight 0. Hence, the cards that remain at the end of the game must also have weight 0. It follows that we can't have three cards left unless they form a set.

We now extend this simple observation to describe the EndGame.

- At the beginning of the game, remove one card C from the deck without looking at it.
- Deal twelve cards face up, and play the game as usual, removing three cards that form a set and replacing them.
- At the end of the game, let S be the collection of cards remaining face up, not including the hidden card.
- Then the hidden card C is the unique card that makes $w(S \cup C) = 0$.
- Finally, now that you've determined the card C, you might be lucky enough to find a set included in $S \cup C$. Note that such a set, if it exists, must include C.

In a simulation that played the game ten million times, Brian Lynch (Lafayette College, Easton, Pennsylvania, 2013) found that the hidden card forms a set with two of the remaining cards just over one-third of the time. When this happens, it can have the flavor of a magic trick.

In practice, a few different procedures can be used to figure out the hidden card without resorting to pencil and paper. We describe two procedures that we explain via an example. Consider the eight cards shown in Figure 14.2.

In the first procedure, mentally remove "sets," considering only one attribute at a time.

- Number: For the cards left in Figure 14.2, you can make one 1–2–3 "set" and one 3–3–3 "set." This leaves 1–2, so the hidden card has to have three symbols.

Figure 14.3. The missing card.

- Color: As before, remove "sets," this time using only color: We can remove a red–red–red "set" and a green–green–green "set," leaving us with purple–purple. That means the hidden card is purple.
- Do the same thing for shading and shape.

In this case, the missing card is the one shown in Figure 14.3.

It doesn't matter how you group the cards into "sets" as you do this. For instance, in figuring out the number for the hidden card, we could have grouped the cards as follows: Removing two 1–2–3 "sets" leaves 3–3, so we find the hidden card's number is 3, which agrees with our previous calculation.

This procedure can be challenging to keep track of in your head. One way to speed this up is to try to determine two or more attributes at once. For example, the three cards in the bottom row on the left of the figure are three green striped squiggles, three purple striped diamonds, and three red solid ovals.

These three cards form a "set" if you ignore shading. If you first determine the shading of C as described above, you can then eliminate these three cards from your consideration for the three remaining attributes. The entire process is significantly easier with five cards than with eight, and it can be quite challenging with eleven cards.

An alternate procedure also works attribute-by-attribute but is occasionally easier to implement.

- Number: For the cards left in Figure 14.2, let's count the number of 1s, 2s, and 3s. In this case, we find two 1s, two 2s, and four 3s. We need these three numbers to be the same modulo 3, so there must be five 3s, including C. So, we determine the number for the hidden card is 3.
- Color: We have three reds, three greens, and two purples. So we need another purple to make these three numbers equal modulo 3, which forces the missing color to be purple.
- Do the same thing for the remaining attributes.

In this procedure, for each attribute, we just make sure that each value is represented the same number of times modulo 3. It is not hard to show why this procedure works.

Finally, does the hidden card form a set with the other cards? Sometimes. In the example shown in Figure 14.2, the hidden card is actually in two different sets with the remaining cards.

Figure 14.4. Two non-sets taken during play.

2 Error Detection (and Correction?)

What happens when a player mistakenly takes three cards that are not a set, but no one notices? Is there a way to determine that an error was made in the play of the game by examining the cards remaining at the end of the game? If we find the error, can we correct it? Hamming weights and the EndGame help us answer these questions.

2.1 Error Detection

Let's start with an example that uses our EndGame parity check. Suppose, in playing the EndGame, we determine the missing card is one purple striped oval. But when we turn over the hidden card, we get a surprise: the hidden card has two purple striped ovals instead. What does this information tell us?

First, from our analysis of the EndGame, we know that at least one mistake was made over the course of the game (i.e., at least one non-set was taken). But we can say more in this case. Since the missing card was incorrect in number, then, if only one incorrect set was taken, the mistake must have been in number.

However, if more than one mistake was made, it is possible that some of those mistakes canceled each other out. For example, suppose that the two non-sets pictured in Figure 14.4 were taken during the game.

In this case, the non-set on the left has a mistake in shading, and that on the right has a mistake in number and shading, but the two mistakes in shading cancel each other out, so the errors will be undetectable from the EndGame. This means the expected and actual cards in the EndGame will differ only in number.

What if two mistakes are detected in the EndGame? For instance, suppose we expected to get three red striped squiggles, but when the hidden card is turned over, it has three green empty squiggles. In this case, our prediction is wrong in both color and shading. If the players then look through their sets in search of the non-sets, and one player finds a non-set whose mistake was only in color, then there must have been at least one other non-set taken during the game that contained a mistake in shading.

In practice, mistakes involving more than one attribute are rarely made, and almost never occur among experienced players. We'll call a mistake *bad* if it involves taking three cards that have a mistake in more than one attribute.

Figure 14.5. A bad non-set.

We now describe error detection using the EndGame. Let C be the hidden card from the EndGame, and let S be the collection of cards remaining face up at the end of the game, not including the card C. Let C' be the card predicted by the EndGame. Then it is easy to verify both of the following.

1. If no errors were made, then $C = C'$ and $w(S \cup C) = 0$.
2. If $C \neq C'$, then at least one error was made. Assume no bad errors were made in the play of the game (i.e., all non-sets taken were wrong in exactly one attribute). Then the number of errors made in the play of the game is at least $w(S \cup C)$.

If $w(S \cup C) > 1$, then either more than one non-set was taken during the game, or at least one bad mistake was made, or both. We can't distinguish these occurrences from the cards $S \cup C$, but when we find the first non-set, we may be able to resolve this uncertainty. But if we assume at most one non-set was taken (a common assumption in coding theory), then $w(S \cup C)$ tells us precisely how bad that mistake was.

2.2 Error Correction?

Suppose we detect an error using the EndGame. Let's assume that exactly one error was made in the play of the game, and it was detected using our EndGame parity check. The players then look through their piles of cards, and one player finds the non-set shown in Figure 14.5. The coordinates for these three cards are $(1, 0, 1, 1)$, $(1, 2, 0, 2)$, and $(2, 1, 1, 0)$. What is the Hamming weight of this non-set? Summing the coordinates modulo 3, we get $(1, 0, 2, 0)$, with Hamming weight 2. Since this vector sum is nonzero in the first and third coordinates—the coordinates that correspond to number and shading—these are precisely the attributes for which an error was made.

Was there a wrong card in the non-set? Can we somehow correct the error that was made? Here is the procedure that we can use to address these questions.

- Let $\{A, B, C\}$ be a non-set. In our example (Figure 5), we have $A = (1, 0, 1, 1)$, $B = (1, 2, 0, 2)$, and $C = (2, 1, 1, 0)$.
- Compute the vector sum $A + B + C$. For the example, we get $(1, 0, 2, 0)$.
- Write $E = -(A + B + C) \pmod 3$. In the example, we have $E = (2, 0, 1, 0)$. We call the vector E the *error vector*.

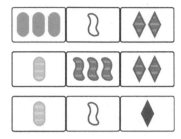

Figure 14.6. The non-set from Figure 14.5 fixed three different ways. Note that each row forms a set.

- Then $A + B + C + E = (0, 0, 0, 0)$, so we can add E to any of A, B, or C to produce a set and fix the error. We list the coordinates for the sets here; see Figure 14.6 for the corresponding cards:

  1. $\{(A + E), B, C\} = \{(0, 0, 2, 1), (1, 2, 0, 2), (2, 1, 1, 0)\}$,
  2. $\{A, (B + E), C\} = \{(1, 0, 1, 1), (0, 2, 1, 2), (2, 1, 1, 0)\}$, and
  3. $\{A, B, (C + E)\} = \{(1, 0, 1, 1), (1, 2, 0, 2), (1, 1, 2, 0)\}$.

Given any two SET® cards, there is a unique third card that completes a set. Hence, the three sets pictured in Figure 14.6 could be found by simply taking two of the three cards in our non-set and finding the card that completes the set. We conclude that it is impossible to say which card was "wrong." However, we have found three ways to correct the mistake. We summarize this result with a proposition.

Proposition 2. *Suppose $\{A, B, C\}$ is a non-set, and let $E = -(A + B + C)$, where this vector sum is computed modulo 3. Then $\{(A + E), B, C\}$, $\{A, (B + E), C\}$, and $\{A, B, (C + E)\}$ are all sets.*

Hence, error correction is not unique, even when only one non-set was taken and that non-set was wrong in only one attribute.

3 Perfect Codes

Coding Theory is a very important area of mathematics, with applications found daily in your cell phone, your iPod®, your unmanned spacecraft, and virtually every situation where data are transmitted. We assume the reader has no background in coding theory. For a quick introduction, we recommend Sarah Spence Adams's short monograph [1]. A more advanced treatment can be found in van Lint's graduate-level textbook [6].

For our purposes, a *linear code* is a collection of cards whose corresponding vectors are closed under addition and scalar multiplication. This is the definition of a *subspace* in linear algebra. This restriction forces the card corresponding to the 0-vector, namely, three green empty diamonds, to be in our code. (This card is arbitrary, of course, reflecting the arbitrary choices we made in assigning coordinates.) We relax this restriction later in this section.

We think of the elements of the linear code as *codewords* that will be transmitted as data from a source. When a codeword is sent, however, some error may occur. For instance, one bit of a transmission may be misread, or coded incorrectly, or changed by a random cosmic ray. We have a problem: we have received a vector as a transmission, but that vector does not correspond to any of our special codewords.

The goal is then to choose the codewords intelligently, so that when the received vector does not match any of them, we can still figure out what word was sent. This allows us not only to detect an error, but also to correct it uniquely, unlike the situation discussed in Section 2.

We need a few definitions.

Definition 2. Let A and B be SET® cards. The *Hamming distance* $d(A, B)$ is the number of attributes for which the cards differ.

One can check that this definition of distance agrees with our *Hamming weight* from Definition 1 in Section 1. So if v_A and v_B are the vectors that correspond to the respective cards A and B, then we have

$$d(A, B) = w(v_A - v_B).$$

The next definition is also adapted from coding theory.

Definition 3. The *Hamming ball* $B_r(A)$ of radius r about the SET® card A is the collection of all cards of distance at most r from A:

$$B_r(A) = \{C \mid d(A, C) \le r\}.$$

When using the code, our source will send a codeword. If we receive that codeword, we assume no errors were made and decode it with no changes. If our received word does not match one of the codewords, though, then an error must have crept in. To fix this, we choose the *closest* codeword, using the concept of Hamming distance.

To have the code actually work, we want the Hamming balls of radius r to have three properties:

1. The centers of all the balls are the cards corresponding to codewords;
2. No two balls intersect;
3. Every one of the eighty-one cards is in exactly one of the Hamming balls.

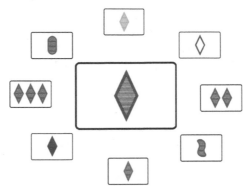

Figure 14.7. The ball of radius 1 about a SET® card.

TABLE 14.2.
Counting the number of cards of distance k from a given card

Distance k	0	1	2	3	4
Number of cards of distance exactly k	1	8	24	32	16

TABLE 14.3.
Counting the number of cards in $B_r(A)$

k	0	1	2	3	4
$|B_r(A)|$	1	9	33	65	81

These conditions ensure that every possible received card has a unique closest codeword. We need to answer a few counting questions to find the radius r that works.

- How many cards are in the Hamming ball $B_r(A)$ for $0 \leq r \leq 4$?
- For what values of r can you completely partition the deck into disjoint balls of radius r?

To answer the first question, first choose the $k \leq 4$ attributes that do not match the card A. Now, we have two choices for each of those attributes, so there are precisely $\binom{4}{k}2^k$ cards of distance k from A. See Table 14.2.

To find the number of cards in the Hamming ball of radius r, we need a sum: $|B_r(A)| = \sum_{k=0}^{r} \binom{4}{k}2^k$. We keep track of these numbers in Table 14.3.

To partition the entire deck of eighty-one cards into disjoint balls, we require that the number of cards in a Hamming ball divides eighty-one (i.e., $r = 0, 1$, or 4). This gives us three potential codes, but two of these are trivial. Of course, you can partition the deck into eighty-one balls of radius 0, or you can partition the deck into one ball of radius 4, but those solutions are not interesting. But choosing $r = 1$ will give rise to our perfect linear code.

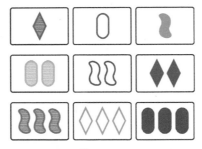

Figure 14.8. A code composed of SET® cards. Note $d(X, Y) = 3$ for any two cards in the subspace.

Note that there are nine cards in $B_1(A)$, the Hamming ball of radius 1 about the card A. But there are also nine cards in a two-dimensional linear code; this is a two-dimensional subspace. Since $9 \cdot 9 = 81$ exhausts the deck, it should be possible to find nine cards so that the following properties hold:

1. The nine cards form a *subspace* (i.e., are closed under addition and scalar multiplication); and
2. The nine balls of radius 1, centered about the nine codeword cards, are *pairwise disjoint* (i.e., $B_1(A) \cap B_1(C) = \emptyset$ for any distinct cards A and C in our subspace).

To find one of these subspaces, note that we need $d(A, C) \geq 3$ for any two cards in the code: otherwise, the balls of radius 1 about those two cards will intersect. It is an interesting fact that if you choose any three cards that aren't a set, where each of the three pairs has distance 3, then the subspace containing those three cards must have $d(X, Y) = 3$ for *every* pair of cards X and Y in the entire subspace. What this means for SET® cards is that every pair of cards in the subspace share only one attribute (Figure 14.8).

We now relax our dependence on the arbitrary assignment of coordinates. Translating any such two-dimensional subspace by adding a fixed vector to every vector in that subspace does not change property 2 above. We call such a translated subspace a *code plane*. In ordinary Euclidean geometry, planes don't need to contain $\vec{0}$, and in affine geometry, these translates are planes as well.

What does the corresponding partition of the deck into disjoint Hamming balls look like? One such partition is shown in Figure 14.9. Several interesting patterns are associated with this partition, and we encourage interested readers to search for them. In particular, the cards are organized so that the collection of nine cards in a certain location in each ball also forms a code plane. Thus, for example, the reader can verify that the nine cards directly above the nine centers also form a code plane (this plane is a translation of the subspace formed by the cards in Figure 14.8).

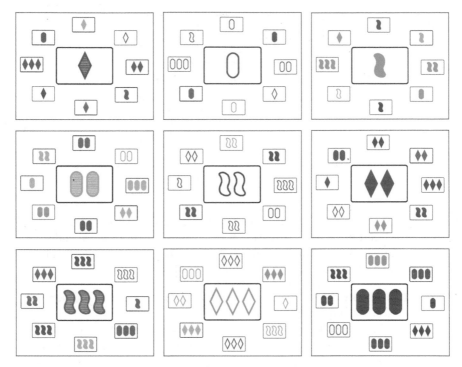

Figure 14.9. A partition of the SET® deck into disjoint Hamming balls of radius 1.

Finally, how does this partition function as a code? Imagine you wish to send a message, and that message is one of the nine cards in the code plane of Figure 14.8. For example, you wish to transmit two green striped ovals from the Mars Rover *Curiosity*. But an error occurs, say, in color, and the message received is two red striped ovals. Using the partition of the deck into disjoint Hamming balls of Figure 14.9, you can correctly decode to the center of the ball containing the received card, namely, two green striped ovals.

Summing up, we have a *perfect, single-error-correcting code. Perfect* simply means that every vector is covered by the disjoint balls, and *single-error correcting* means that any mistake in a single attribute can be corrected uniquely.

4 Further Exploration

We conclude by offering a few counting problems that interested readers may wish to explore on their own. Many of these results rely on the geometry of the SET® deck [4].

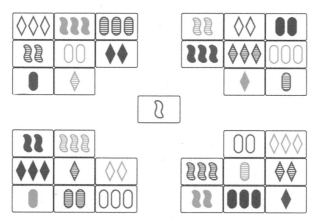

Figure 14.10. All cards are distance 3 from the card C in the center; the four planes all include C.

The total number of planes in the SET® deck is 1,170; 130 planes contain a given card.

1. How many of the 1,170 planes in the deck are code planes?
2. Fix a card. How many of the 130 planes that contain that card are code planes?

You can verify that the answers to these two questions are 72 and 8, respectively. One consequence: the probability that a randomly chosen plane is a code plane is $72/1{,}170 \approx 6\%$. Note that the global and local probabilities are the same: $72/1{,}170 = 8/130$.

We examine the eight code planes containing a given card C in more detail; for now, let C be the card with one purple empty squiggle. Recall from Table 14.2 that there are exactly thirty-two cards of distance 3 from C. It turns out that you can partition these thirty-two cards into four disjoint sets of size eight, where each of these form a code plane with C. And you can do this in two different ways. Figure 14.10 shows one example.

Now let D be the card with one green striped diamond. Then D is in one plane in Figure 14.10, but it will also be in one other code plane with C. That plane is shown in Figure 14.11. You can verify that, when you start with either of these planes, there is only one way to decompose the remaining cards into three disjoint planes containing C. An examination of the incidences among the points, lines, and planes in this context should reveal additional structural properties.

Additionally, if we start with two cards C_1 and C_2 with $d(C_1, C_2) = 3$, the spheres of radius 3 about C_1 and C_2 intersect in thirteen cards.

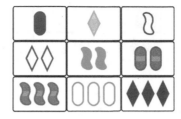

Figure 14.11. One more plane with C and the card with one green striped diamond.

Furthermore, each card of weight 3 is orthogonal to itself, since its dot product with itself is 0. Thus each of the code planes is self-dual. It would be interesting to investigate this property in other dimensions.

More generally, much of what we've done here can be extended to higher dimensions. Other perfect ternary codes have interpretations through higher-dimensional decks. In particular, perfect ternary codes exist in dimensions 4, 11, 13, and 40, and no other dimensions less than 100. We do not recommend actually playing the eleven-attribute Set game with $3^{11} = 177{,}147$ cards, however.

References

[1] S. S. Adams. Introduction to Algebraic Coding Theory. http://www.math. niu.edu/~beachy/courses/523/coding_theory.pdf (accessed April 14, 2015).

[2] B. Davis and D. Maclagan. The card game SET®. *Math. Intelligencer* **25** no. 3 (2003) 33–40.

[3] M. Follett, K. Kalail, E. McMahon, and C. Pelland. Partitions of $AG(4, 3)$ into maximal caps, arXiv:1302.4703.

[4] H. Gordon, R. Gordon, and E. McMahon. Hands-on SET®, *PRIMUS: Problems, Resources, and Issues in Mathematics Undergraduate Studies* **23** no. 7 (2013) 646– 658.

[5] SET® Enterprises. http://www.setgame.com/ (accessed April 14, 2015).

[6] J. H. van Lint. *Introduction to Coding Theory*, third edition. *Graduate Texts in Mathematics* 86. Springer-Verlag, Berlin, 1999.

15

CONNECTION GAMES AND SPERNER'S LEMMA

David Molnar

Connection games are a family of abstract strategy games for two players. The best-known is Hex, independently invented by Piet Hein and John Nash in the 1940s. Martin Gardner is at least partially responsible for the popularity of these games, having written about Hex early in his career [7]. The genre has exploded in the past decade; an extensive taxonomy appears in Browne [4].

After providing an introduction to connection games in general, I recount how Sperner's Lemma, a result about labeling a triangulation of a simplex, can be used to prove that someone must win at Hex, as well as The Game of $Y^®$, (or simply, Y) another well-known connection game.

In early 2008, Mark Steere published two new connection games, Atoll and Begird [10, 11], which can be played on a variety of boards and include Hex and Y, respectively, as special cases. I prove a generalization of Sperner's Lemma and use it to show that there is always a winner in the many variations of Atoll and Begird. These "must-win" results have significant strategic implications— if you prevent your opponent from making the desired connection, you will make this connection yourself by necessity!

1 Hex and Y

The Hex board is a quadrilateral tessellated by hexagonal cells. Each player owns a pair of opposite sides and aims to create a chain of cells in his or her own color connecting these two sides (see Figure 15.1a). In Y both players aim to connect all three sides of a triangular board (see Figure 15.1b) The Game of Y was invented in 1953 by Charles Titus and Craige Schensted (who has since changed his name to Ea Ea).

Connection games characteristically can be played in equivalent, dual forms. With pencil and paper, it is preferable to play on the cells—players take turns coloring the cells in with the players' assigned colors, no special equipment being needed. Mathematically, however, it is convenient to view games as taking place on the vertices. Let us borrow the concept of a *dual graph*

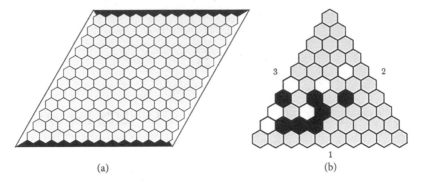

Figure 15.1. Game boards for (a) Hex and (b) Y.

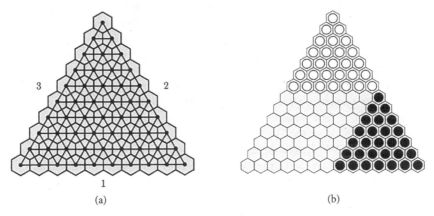

Figure 15.2. (a) Construction of the dual graph. (b) Hex embedded in Y.

from map-coloring problems. Given a game board composed of cells, the dual graph is constructed by placing a point inside each cell and connecting two points with an edge when the corresponding cells are adjacent (Figure 15.2a). This process is referred to as placing black and white *stones* on the vertices. Pencil-and-paper boards generally use a hexagonal array, but it is not the shapes, but rather the connectivity, of the cells that matters; all that is necessary to prevent deadlocks is that cells meet only in threes. Schensted and Titus call this the *mudcrack principle* [9]. Equivalently, the dual graph is composed entirely of triangles (Figure 15.2b). The following section demonstrates the importance of this condition.

2 Someone Must Win

While it may seem intuitively clear that someone must win at Hex, that is certainly not the case for Y. I learned a beautiful proof that someone must win at Y from Robert Hochberg. This result was subsequently published in joint work with McDiarmid, and Saks [8]. The proof relies on the following combinatorial result.

Lemma 1 (Sperner). *If a triangle is subdivided into smaller triangles in an edge-to-edge fashion, and each vertex is labeled 1, 2, or 3 so that no vertex labeled i appears on side i of the large triangle, where i = 1, 2, 3, then some small triangle bears all three labels.*

A proof can be found in Aigner and Zeigler [1], which also discusses the relationship between the lemma and Brouwer's Fixed Point Theorem.

Theorem 1. *Someone must win at Y.*

Proof. The board for Y in its dual form is a triangle subdivided into smaller triangles. Assume for a contradiction that the board has been filled with white and black stones, but that neither player's stones connect all three sides. It is not immediately obvious how to construct the Sperner labeling, as there are three possible labels and only two colors of stones. There are, however, three sides, and so we label each vertex according to the lowest-numbered side that the stone on that vertex is *not* connected to. By hypothesis, this labeling is well defined. Moreover, since a stone on side *i* is certainly connected to side *i*, this process defines a Sperner labeling. By Sperner's Lemma, there is a small triangle with all three labels. But two of the stones on the three vertices of this small triangle are necessarily of the same color. By construction, adjacent stones of the same color cannot have different labels, a contradiction. □

It is then a short step to see why someone must win at Hex [4]. A Hex board can be extended to a Y board with some stones already placed on it (see Figure 15.2b). Someone must win this Y game, and a winning group on the Y game necessarily contains a winning chain on the Hex board for the same player.

3 Star Y

Hex and Y belong to the category of *absolute path* games [4]. Every stone belongs to a unique connected *group* in its color. In an absolute path game, the goal is to create a single group having certain properties whose existence automatically prevents the opponent from creating such a group. We call such a group a *winning group*. Not all connection games have goals defined in terms of a single group.

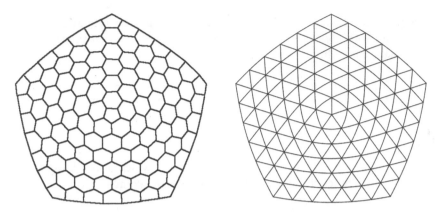

Figure 15.3. A Star Y board and its dual.

When designing new connection games, ideally one is looking for goals that are mutually exclusive, yet a win is guaranteed for one player or the other. Consider a possible connection game on a board with five sides. If the goal were to connect four sides, the game would likely end with neither player winning. In contrast, a goal of connecting three sides is too easy; both players could work toward the goal in separate areas of the board, with no interaction. Between three and four, we find the game Star Y, as described on Ea Ea's website: the goal is to connect any side with both opposite sides [6] (see Figure 15.3). (Other sides may be connected as well.) As we shall see, the players' goals are complementary: exactly one player *must* make such a connection.

To prove this, let us reformulate the winning condition in a way more suitable for proof by contradiction: a player wins by creating a group that *does not miss any two adjacent sides of the board*. Now we need a tool that, applied to pentagons (or other polygons), gives a conclusion similar to Sperner's Lemma. Some care is needed to define "pentagon", or indeed, *n*-gon—we could, for example, play either Hex or Y on a map of the United States, depending on how we defined the sides of the board.

4 What is a Polygon?

The proof of Theorem 1 is purely topological: the size and shape of the board are immaterial. Our definition of polygons retains this generality. Here, and in what follows, name the sides of a board 1 through n, so side numbers (and vertex labels) are elements of $[n] = \{1, \ldots, n\}$. When considering addition or subtraction modulo n, it is this set, rather than $\{0, \ldots, n - 1\}$, that I have in mind.

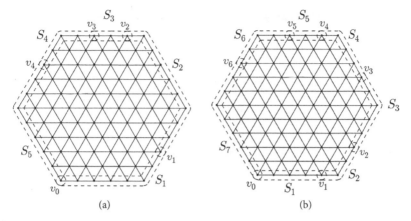

Figure 15.4. (a) A triangulated 5-gon and (b) 7-gon.

Define an *n-gon*, for any $n \geq 3$, to be an embedded planar graph Γ along with n distinguished sets of vertices $S_1, \ldots, S_n$ (the *sides*), so that for each $i \in [n]$,

- $\cup_{i=1}^{n} S_i$ is the set of all vertices incident with the unbounded face of the graph;
- every edge incident to the unbounded face of the graph connects two vertices in some S_i;
- $S_i \cap S_{i+1} \pmod{n}$ is a single vertex, denoted by v_i (except that $S_n \cap S_1 = \{v_0\}$); and
- $S_i \cap S_j$ is empty unless $j = i - 1, i,$ or $i + 1 \pmod{n}$.

Refer to the S_i as the *sides* of the *n*-gon to avoid confusion with the edges of the graph; similarly, the points of intersection v_i are the *corners* of the *n*-gon, rather than its vertices. Note that side S_i stretches from v_{i-1} to $v_i, i = 1, \ldots, n$. The sides are defined implicitly by the corners (Figure 15.4.) When all interior faces of Γ are triangles, we say that Γ is a *triangulated n-gon*.

For the games under consideration, the term *n-sided board* refers to a triangulated *n*-gon.

The following result is proved in greater generality later.

Lemma 2. *Let Γ be a triangulated 5-gon. A labeling of the vertices of Γ with labels in [5] such that the vertices of side S_i are not labeled i or $i - 1$ (mod 5) contains a triangle with three distinct labels.*

This lemma implies that someone must win in Star Y. To see this, imagine that a Star Y board has been completely filled in without either player having

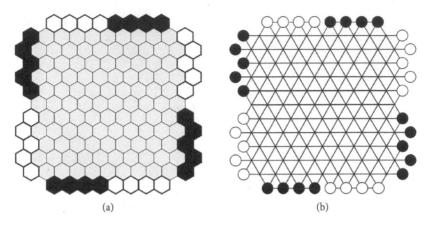

Figure 15.5. Standard Atoll and its dual.

created a winning group. We can label each vertex of Γ with the smallest value of i such that the stone on that vertex is *not* connected to side i or side $i + 1$. Vertices on side i will not be labeled i or $i - 1$. So, by Lemma 2, some triangle has three distinct labels. This is impossible, since there are only two colors of stones, and so contradicts the assumption that neither player has a winning group.

An appealing feature of connection games is the application of metarules that alter game play [4]. For example, the so-called *master rule* addresses the issue of first-player advantage by having the first player place one stone, while on subsequent moves two stones are placed. Alternatively, the *pie rule* states that after the very first move of the game, the second player may choose to swap colors, so a move by the first player that is too good will end up working against him or her. Proofs such as the one given for Theorem 1 respect this mutability: any number of stones can be placed on the board before the game starts, and there still must be a winning connection. Nor is there, in these proofs, any assumption about the number of white vs. black stones, or that the players alternate turns, or any other formulation of fairness.

5 Atoll

The game *Atoll* uses the familiar hexagonal grid, but it relaxes the limitations on the number of sides by introducing islands along the perimeter [10]. A standard board features eight islands of alternating color (Figure 15.5). The goal is to connect either pair of opposite islands of your color. A winning connection may pass through the cells of an intervening island.

To dualize Atoll and retain the island structure, we consider the cells comprising the islands as corresponding to vertices in the dual graph with

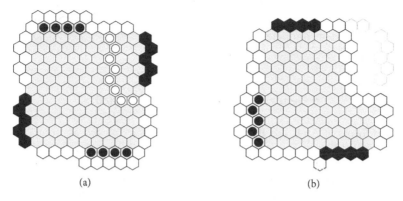

(a)   (b)

Figure 15.6. Proof that someone must win at Atoll.

stones already placed on them. Observe that pairs of neighboring islands share exactly one unoccupied cell to which they are both adjacent; the corresponding vertices in the dual graph constitute the corners. The islands and the corresponding corners give an 8-gon according to our definition. Despite a squarish appearance, Atoll truly is played on an 8-sided board.

Theorem 2. *Someone must win at Atoll.*

Proof. Let the islands of the Atoll board be numbered counterclockwise, with the odd-numbered islands white and the even-numbered ones black. Extend the white islands so that islands 1 and 3 meet, as well as 5 and 7, and pop islands 2 and 6 onto the board as groups of black stones, as shown in Figure 15.6a,b. Now we have a board with only four islands with alternating colors, which is Hex. Someone must win this game of Hex. If the winner is Black, then Black has connected islands 4 and 8 of the Atoll board, and we are done with this proof. Note that this connection may pass through the popped stones, showing why allowing a winning connection to pass through an island is necessary.

If the winner of the Hex game is White, then it could be that White has connected islands 1 and 5, or 3 and 7; in either case, we are done. Otherwise, White must have connected 1 to 7 or 3 to 5. Assume the latter, noting that if it were the former, we could just renumber the sides. Now grab hold of island 4 and break off the part of the board that is cut off by White's 3 to 5 connection. Consider that connection, along with any remaining pieces of islands 3 and 5, to be a single white island. Revert the popped black stones of islands 2 and 6 to island status. Extend islands 1 and 7 so that they meet, popping the black island 8 onto the board. Again, we have a Hex game. If Black wins, then Black has connected islands 2 and 6, winning the Atoll game. If White wins, then White has connected either island 1 or island 7 to the already-existing connection between 3 and 5. In either case, White has made a winning connection in the Atoll game, which completes the proof. $\square$

6 General Forms of Atoll and Begird

General forms of Atoll and Begird (defined below) are played on boards with many sides and have more complicated winning conditions. For example, on a twelve-sided board, connecting a pair of opposite sides, as in standard Atoll, is not an appropriate goal; one player creating a group connecting S_4, S_8, and S_{12} (and no other sides) would prevent a win by the other player without achieving a win.

Now let us define the span of a group in such a way that a player wins by creating a group that precludes the existence of any group of equal or greater span in the other color. Note that merely counting the sides a group touches will not accomplish this. Instead, define the *span* of a connected group G on an n-sided board as the smallest k such that the set of sides connected by G is contained in a stretch of k consecutive sides. Equivalently, the span of a group is at least k if the group never misses $n - k + 1$ consecutive sides of the board. (If G touches no sides, its span is 0.)

The winning conditions for Y or Star Y are now seen to be the creation of a group with span 3 or 4, respectively. This suggests the following generalization of those two games. The game Begird(n) is played on an n-sided board, where $n = 2m + 1$, $n \geq 3$. A winning group is one with span at least $m + 2$. With this terminology, we see that Y is Begird(3), and Star Y is Begird(5). (Perhaps constrained by board geometry, Steere [11] only considers Begird games where the number of islands is an odd multiple of three and names Begird(15) as the "standard" form.)

Atoll(n) is played on an n-sided board, where n is a multiple of four, all S_i have cardinality at least 3, and stones have already been placed on the noncorner vertices of S_i (black for odd i and white for even i). A winning group is one with span at least $(n + 2)/2$. Atoll(4) is simply Hex, and Atoll(8) is the standard version described above. Note that for $n = 8$, a group has span at least 5 if and only if it connects two opposite islands. See Figures 15.3 and 15.5 for examples.

We could try using the "popping" idea from Theorem 2 to construct an induction argument that someone must win at all versions of Atoll. However, it is more attractive to ask whether it is possible to reverse-engineer a result similar to Sperner's Lemma that could be used to show that someone must win at all forms of Begird, and then obtain the result for Atoll as a corollary. Since the goal of Begird($2m+1$) is to not miss any m consecutive sides, what we want is to eliminate m consecutive labels along each side of the $(2m + 1)$-gon.

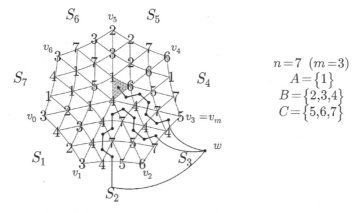

Figure 15.7. Relaxed Sperner labeling of a heptagon and its partial dual.

7 Extending Sperner's Lemma

Define a *relaxed Sperner labeling* of a triangulated n-gon $\Gamma = (V, E, \{S_i\}_{i=1}^n)$ to be a function $\lambda : V \to [n]$ such that if $v \in S_i$, then

$$\lambda(v) \in \{i + 1, \ldots, i + m + 1\} \pmod{n},$$

where $m = \lfloor (n - 1)/2 \rfloor$. For $n = 3$, this is the usual Sperner labeling of a triangle. As with the usual Sperner labeling, this condition rules out certain labels along each side of the n-gon, but not in the interior. Figure 15.7 shows an example, where vertices on side 1 have labels in $\{2, 3, 4, 5\}$, vertices on side 5 have labels in $\{6, 7, 1, 2\}$, and so forth.

Theorem 3 (Relaxed Sperner's Lemma). *Let Γ be a triangulated n-gon with vertices V and sides $S_1, \ldots, S_n$. If λ is a relaxed Sperner labeling on Γ, then Γ contains a triangle with three distinct labels.*

Note that the usual Sperner's Lemma is the case $n = 3$. The relaxed version has a weaker hypothesis than the generalization in Atanassov [3], but a weaker conclusion as well.

Proof. Let $m = \lfloor (n - 1)/2 \rfloor$. For n odd, define

$$A = \{1\}, \quad B = \{2, 3, \ldots, m + 1\}, \quad \text{and} \quad C = \{m + 2, m + 3, \ldots, n\}.$$

For n even, define

$$A = \{1, n\}, \quad B = \{2, 3, \ldots, m + 1\}, \quad \text{and} \quad C = \{m + 2, \ldots, n - 1\}.$$

Construct a partial dual Γ^* to Γ as follows. The vertices V^* of Γ^* correspond to the triangles of Γ, with an extra vertex w representing the external face of Γ. If x and y are vertices in V connected by an edge e and if $\lambda(x) \in B, \lambda(y) \in C$, and $\lambda(y) - \lambda(x) \le m$, then the vertices in V^* corresponding to the faces incident to e are themselves connected by an edge in Γ^*, and we say that e is *crossed*.

I claim that the degree of w is odd, that is, an odd number of boundary edges of Γ are crossed. The argument is as follows. For a relaxed labeling, $\lambda(v_0)$ must be in B, and $\lambda(v_m)$ is in C. Furthermore, there can be no vertices on the sides $S_1, \ldots, S_m$ with label in A. So, as we walk along the sides from v_0 to v_m, the labels of the vertices switch back and forth between B and C an odd number of times. In S_i, for $i = 1, \ldots, m$, the permissible labels in B are $\{i+1, \ldots, m+1\}$, and the permissible labels in C are $\{m+2, \ldots, m+i+1\}$. So along these sides, any edge incident to vertices x and y with $\lambda(x) \in B$ and $\lambda(y) \in C$ automatically satisfies $\lambda(y) - \lambda(x) \le m$. Therefore, an odd number of edges along the sides $S_1, \ldots, S_m$ are crossed.

Along the remaining sides S_i for $i = m+1, \ldots, n$, the permissible labels in B (if there are any) are no greater than $i - m$, and the permissible labels in C (if there are any) are at least $i + 1$. So no edge along these sides can be crossed: when $\lambda(x) \in B$ and $\lambda(y) \in C$, then $\lambda(y) - \lambda(x) \ge m + 1$ must hold. This establishes the claim made in the previous paragraph.

The vertices in V^* other than w have degree 0, 1, or 2, for if x, y, z are vertices of a triangle, there is no way for the pairs (x, y), (y, z), and (x, z) all to have one vertex labeled in B and the other in C.

Since the sum of the degrees of all vertices of Γ^* is even, some vertex in $V^* \setminus \{w\}$ has degree 1. This vertex corresponds to a triangle in Γ with vertices x, y, z, with, say, the edge incident to x and y crossed. This implies $\lambda(x) \ne \lambda(y)$. But if $\lambda(z) = \lambda(x)$, then the edge incident to y and z would also be crossed, and if $\lambda(z) = \lambda(y)$, then the edge incident to x and z would also be crossed. Therefore x, y, and z all have different labels. $\qquad\square$

8 Once Again—Someone Must Win

In this section we shall prove that draws are impossible in both Begird and Atoll in their general formulations.

Theorem 4. *Someone must win at Begird(n), for any odd $n \ge 3$.*

Proof. Recall that if we write $n = 2m + 1$, a winning configuration is a group with span greater than or equal to $m + 2$. Suppose, contrary to what we want to prove, that for some Begird(n) board, Γ, the game has been played to conclusion without a winner. That means a stone has been placed on every vertex of Γ and that no group has span greater than or equal to $m + 2$. Using the alternate characterization of span from Section 6, each group must somewhere miss $(2m + 1) - (m + 2) + 1 = m$ consecutive sides. Thus, for each stone, there exists some $k \in [n]$ such that the stone is not connected to sides $S_k, S_{k+1}, \ldots, S_{k+m-1} \pmod{n}$. Accordingly, define $\lambda : V \to [n]$, where $\lambda(v)$ is the least such k for the stone on v. The claim is that this is a relaxed Sperner labeling. If $v \in S_i$, then the stone on v is certainly connected to side S_i, so that

$$\lambda(v) \in [n] \setminus \{i - m + 1, \ldots, i\} = \{i + 1, \ldots, i + m + 1\} \pmod{n}.$$

The Relaxed Sperner Lemma then ensures that some triangle in Γ has three distinct labels under λ. But two of the stones on these vertices are the same color, meaning they are part of the same group. So these vertices must have the same label. This is a contradiction, so some group must have span at least $m + 2$. This proves a draw is impossible in Begird. $\qquad\square$

Theorem 5. *Someone must win at Atoll$(2m + 2)$, for m odd and $m \geq 1$.*

Proof. Recall that Atoll(n) is played on an n-gon board, where n is divisible by 4. Now write $n = 2m + 2$, where m must be odd to ensure that n is evenly even. The idea of the proof is to embed the Atoll$(2m+2)$ board in a Begird$(2m+1)$ board, much as in Figure 15.2, where a Hex board (i.e., Atoll(4), $m = 1$) is embedded in a Y board (i.e., Begird(3)).

Given a filled-in Atoll$(2m + 2)$ board with sides $A_1, \ldots, A_{2m+2}$, we need to show that there is some group G that does not miss any $m + 1$ consecutive sides, that is,

(1)   for all $k \in [2m + 2]$, there exists an $i \in \{k, k + 1, \ldots, k + m\} \pmod{n}$
such that G is connected to A_i.

Now, as embedded graphs, the Atoll board and the containing Begird board can be the same. All that needs to be done is to define the sides of the Begird board, $\{B_i\}$. Take $b_0 = A_n \cap A_1$. For each i except 1 and n, select a noncorner vertex $b_{i-1} \in A_i$. These become the corners of the Begird board; that is, B_i is that part of the perimeter of the board from b_{i-1} to b_i, as shown in Figure 15.8. Note that as we are defining different sets of sides for the same board, we need to distinguish between two possibly different spans, Atoll span and Begird span.

Since someone must win at Begird, there is some group of stones G with Begird span at least $m + 2$. Therefore, G does not miss any m consecutive Begird sides. That is,

(2)   for all $j \in [2m + 1]$, there exists some
$i \in \{j, j + 1, \ldots, j + m - 1\} \pmod{(n - 1)}$ such that G is connected to B_i.

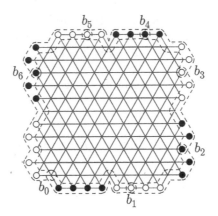

Figure 15.8. Atoll(8) board sitting inside a Begird(7) board.

I claim that statement (2) implies statement (1). Suppose the winning group is black, so consider only the odd-indexed sides A_i. We will define

$$f : [2m + 1] \rightarrow \{1, 3, \ldots, 2m + 1\},$$

so that when a black group is connected to B_i, it is also connected to $A_{f(i)}$. By construction, for each $i \in [2m + 1]$, B_i includes parts of A_i and A_{i+1}. So a black group connected to B_i must be connected to A_i or A_{i+1}, whichever is odd. Thus, define

$$f(1) = 1, \qquad f(2) = f(3) = 3, \qquad f(4) = f(5) = 5,$$

and so forth (i.e., $f(x) = 2\lfloor \frac{x}{2} \rfloor + 1$). However, if instead we define $f(x) = 2\lceil \frac{m+1}{2m+1} x \rceil - 1$, we get a map satisfying

$$f(x + 2m + 1) = f(x) + 2m + 2,$$

which simplifies the remaining analysis.

Given $k \in [2m + 2]$, take $j = k$ if $k \leq 2m + 1$, $j = 1$ if $k = 2m + 2$, and define i as in statement (1). Then G is connected to $A_{f(i)}$ and

$$f(i) \in \{f(j), f(j + 1), \ldots, f(j + m - 1)\} \pmod{(n + 1)}.$$

Note that $f(j) = k$ if k is odd, and $f(j) = k + 1$ if k is even. In either case, since m is odd, $(m - 1)/2$ is an integer, and

$$f(j + m - 1) = 2 \left\lceil \frac{m+1}{2m+1} j + \frac{m-1}{2} - \frac{m-1}{4m+2} \right\rceil - 1$$

$$= 2 \left\lceil \frac{m+1}{2m+1} j - \epsilon \right\rceil + (m - 1) - 1,$$

where $0 < \epsilon = \frac{m-1}{4m+2} < 1$. So

$$f(j + m - 1) = \begin{cases} f(j) + m - 1, & \text{if } \lceil \frac{m+1}{2m+1} j - \epsilon \rceil = \lceil \frac{m+1}{2m+1} j \rceil \\ f(j) + m - 3 \pmod{n}, & \text{otherwise.} \end{cases}$$

We have that

$$\{f(j), \dots, f(j + m - 1)\} \subseteq \{k, \dots, k + m\} \pmod{n}.$$

It follows that $f(i) \in \{k, \dots, k + m\}$; that is, the group G is connected to (at least) one of the sides $A_k, \dots, A_{k+m}$.

Thus the group G never misses $m + 1$ consecutive sides of the Atoll board. Equivalently, the Atoll span of G is at least $n - (m + 1) + 1 = m + 2$; G is a winning group in Atoll as well.

If the winning group G is white, the same result is obtained by reversing the numbering of the sides. □

Remark. When m is even, the argument above gives a somewhat weaker result: a group with minimum span $m + 2$ in Begird$(2m + 1)$ is only guaranteed to not miss more than $m + 1$ consecutive sides of the corresponding Atoll$(2m + 2)$ board; thus its span in Atoll need only be as large as $m + 1$. But, Black and White can simultaneously create groups with this span. If we were to define the goal in this case to be a group of sufficient span to preclude the other player from creating such a group, a win would not be guaranteed.

9 Other Games

When proving that someone must win at Begird, we used the relaxed Sperner's Lemma for odd n. Showing that someone must win at Atoll only used the Lemma for $n = 2m + 2$, where m is even. The remaining case seems to have been wasted. Conveniently, Jorge Gomez Arrausi apparently anticipated this development with the invention of Unlur in 2001 [2]. Unlur is an asymmetric game: on a six-sided board, Black wins by making a Y connecting three alternating sides, and White wins by connecting two opposite sides. There is a "contract phase" that serves to equalize the two goals. These goals are mutually exclusive, but it is possible for neither goal to be achieved. Accordingly, a player who accomplishes the opponent's goal immediately loses, even if accomplishing her own goal simultaneously. It is not the case that one player or the other must have a group connecting three alternating sides. And it is not the case that one player or the other must connect a pair of opposite sides. But one player or the other must make one of these configurations. And so, someone must win.

We can incorporate Unlur into our framework using the concept of span. Note that on a six-sided board, connecting three alternating sides is equivalent to having span at least 5. Moreover, a group with span exactly 4 connects two opposite sides. Thus, showing that someone must win reduces to showing that there must be a group of span at least four.

This observation generalizes. For any $n = 2m \geq 6$, we can define Unlur(n) on a triangulated n-gon, where Black wins by making a group of span at least $m + 2$, and White wins by connecting a pair of opposite sides. In fact, Black's winning condition can be reformulated without reference to span: Black wins by creating a group that makes it impossible for White to connect any pair of opposite sides. The other rules carry over.

Cameron Browne's game Cross, published in 2009 [5], symmetrizes Unlur. Either player wins by connecting three alternating sides and loses by connecting a pair of opposite sides. Cross(n) can be defined analogously. Unlike Unlur, if you achieve both goals simultaneously, you win.

To show, therefore, that someone must win at either Unlur(n) or Cross(n), it suffices to show that one player has a group with span at least $m + 2$. The proof of this fact is virtually identical to the proof that someone must win at Begird. We have then our final result:

for any $n = 2m \geq 6$, someone must win at Unlur(n), and someone must win at Cross(n).

Acknowledgments

I thank Michael Henle and two anonymous reviewers for helpful comments on earlier drafts of this chapter. Paul Zorn and Francis Su provided encouragement early on, as did, without knowing it, the online communities at boardgamegeek.com and yucata.de.

References

[1] M. Aigner, and G. Zeigler, *Proofs from the Book*, fourth edition. Springer, New York, 2010.
[2] J. G. Arrausi, C. J. Ragnarsson, and T. G Seierstad. Unlur. *Abstract Games* **12** (2002) 17–21.

[3] K. T. Atanassov. On Sperner's Lemma. *Studia Sci. Math. Hungar.* **32** (1996) 71–74.

[4] C. Browne. *Connection Games: Variations on a Theme.* A K Peters, Wellesley, MA, 2005.

[5] C. Browne. Cross. http://nestorgames.com/rulebooks/CROSS_EN.pdf (accessed April 15, 2015).

[6] E. Ea. *Star.* http://ea.ea.home.mindspring.com/KadonBooklet/Kadon*Star.html (accessed April 15, 2015).

[7] M. Gardner. The game of Hex. In *Hexaflexagons and the Tower of Hanoi: Martin Gardner's First Book of Mathematical Puzzles and Games*, p. 82. Cambridge University Press, Cambridge, 2008.

[8] R. Hochberg, C. McDiarmid, and M. Saks. On the bandwidth of triangulated triangles. *Discrete Math.* **138** (1995) 261–265.

[9] C. Schensted, and C. Titus. *Mudcrack Y and Poly-Y.* Neo Press, Peaks Island, ME, 1975.

[10] M. Steere. Atoll. http://www.marksteeregames.com/Atoll_rules.pdf (accessed April 14, 2015).

[11] M. Steere. Begird. http://www.marksteeregames.com/Begird_rules.pdf (accessed April 14, 2015).

PART V

◇◇◇◇◇◇◇◇◇◇◇◇◇◇◇◇◇◇◇◇◇◇◇◇◇◇◇◇

Fibonacci Numbers

16

◇◇◇

THE COOKIE MONSTER PROBLEM

Leigh Marie Braswell and Tanya Khovanova

1 The Problem

In 2002, Cookie Monster® appeared in the book *The Inquisitive Problem Solver,* by Vaderlind, Guy, and Larson [3][1]. The hungry monster wants to empty a set of jars filled with various numbers of cookies. On each of his moves, he may choose any subset of jars and take the same number of cookies from each of those jars. The *Cookie Monster number* is the minimum number of moves Cookie Monster must use to empty all the jars. This number depends on the initial distribution of cookies in the jars.

We consider the set of k jars with $s_1 < s_2 < \ldots < s_k$ cookies as a set $S = \{s_1, s_2, \ldots, s_k\}$. We call $s_1, s_2, \ldots, s_k$ a *cookie sequence*. We assume no two s are identical, because jars with equal numbers of cookies may be treated like the same jar. A jar that is emptied or reduced to the number of cookies in another jar is called a *discarded* jar.

Suppose the *Cookie Monster number* of S, which we denote CM(S), is n. On move j, for $j = 1, 2, \ldots, n$, Cookie Monster removes a_j cookies from every jar that belongs to some subset of the jars. We call each a_j a *move amount* of the monster. The set of move amounts is denoted by A. Each jar can be represented as the sum of a subset of A, or as a sum of move amounts.

For example, if $S = \{1, 5, 9, 10\}$, the monster might remove nine cookies from the jars containing nine and ten cookies on his first move, so $a_1 = 9$. After this move, the jar once containing nine cookies is discarded, and the jar once containing ten cookies is reduced to one cookie, a number of cookies in another jar, and is therefore also discarded. On the monster's second move, he may remove one cookie from the jars containing one and five cookies, so $a_2 = 1$. Now the jar once containing one cookie is discarded, and the jar containing five cookies is reduced to four cookies. Finally, the monster can

---

[1] After the Cookie Monster problem first appeared in *The Inquisitive Problem Solver* by Vaderlind, Guy, and Larson, further results were published by Cavers [4], Bernardi and Khovanova [2], and Belzner [1].

empty the jar containing four cookies to finish, so $a_3 = 4$. Since the jars were emptied in three moves, we have that $CM(S) \leq 3$. We see that $A = \{9, 1, 4\}$, and each jar in S may be written as a sum of the elements in A. For example, $s_2 = 5 = 1 + 4 = a_2 + a_3$, and $s_4 = 10 = 9 + 1 = a_1 + a_2$. This notation leads to the following observation.

Lemma 1. *Cookie Monster may perform his optimal moves in any order.*

Proof. Suppose $CM(S) = n$, and Cookie Monster follows an optimal procedure to empty all the cookie jars in S with n moves. After the monster performs move n, all jars are empty. Therefore, each jar $k \in S$ may be represented as $k = \sum_{j \in I_k} a_j$ for some $I_k \subseteq \{1, 2, \ldots, n\}$. By the commutative property of addition, the a_j may be summed in any order for each jar k. Thus Cookie Monster may perform his moves in any order he wishes and still empty the set of jars in n moves. □

We start this chapter with known general algorithms in Section 2 and known bounds in Section 3 for the Cookie Monster number. We explicitly find the Cookie Monster number for jars containing cookies in the Fibonacci, Tribonacci, n-nacci, and Super-n-nacci sequences in Sections 4 and 5. We also construct sequences of k jars such that their Cookie Monster numbers are asymptotically rk, where r is any real number, $0 \leq r \leq 1$, in Section 6.

2 General Algorithms

Vaderlind, Guy, and Larson [3] presented general algorithms Cookie Monster may use to empty the jars, none of which is optimal in all cases.

- *Empty the Most Jars Algorithm*, in which Cookie Monster reduces the number of distinct jars by as many as possible on each of his moves. For example, if $S = \{1, 3, 4, 7\}$, the monster would remove four cookies from the jars containing seven and four cookies or remove three cookies from the jars containing seven and three cookies. In both cases, he would reduce the number of distinct jars by two.
- *Take the Most Cookies Algorithm*, in which Cookie Monster takes as many cookies as possible on each of his moves. For example, if $S = \{4, 7, 14, 19, 20\}$, the monster would remove fourteen cookies from the jars containing twenty, nineteen, and fourteen cookies. In doing so, he would remove a total of forty-two cookies, which is more than if he removed nineteen cookies from the two largest jars, seven cookies from the four largest jars, or four cookies from the five largest jars.

- *Binary Algorithm*, in which Cookie Monster removes 2^m cookies from all jars that contain at least 2^m cookies for m as large as possible on each of his moves. For example, if $S = \{8, 9, 16, 18\}$, the monster would remove $2^4 = 16$ cookies from the jars containing 18 and 16 cookies.

3 Established Bounds

There are natural lower and upper bounds for the Cookie Monster number (see Bernardi and Khovanova [2] and Cavers [4]).

Theorem 1. *Let S be a set of k jars. Then we have*

$$\lfloor \log_2 k \rfloor + 1 \leq \mathrm{CM}(S) \leq k.$$

Proof. For the upper bound, Cookie Monster may always empty the ith jar on his ith move and finish emptying the k jars in k steps. For the lower bound, remember that the k jars contain distinct numbers of cookies. Let $f(k)$ be the number of distinct nonempty jars after the first move of Cookie Monster. We claim that $k \leq 2f(k) + 1$. Indeed, after the first move, there will be at least $k - 1$ nonempty jars, but there cannot be three identical nonempty jars. Thus the number of jars plus one cannot decrease faster than twice each time. □

We can show that both bounds are reached for some sets of jars [2].

Lemma 2. *Let S be a set of k jars with $s_1 < s_2 < \ldots < s_k$ cookies. If $s_i > \Sigma_{k=1}^{i-1} s_k$ for any $i > 1$, then $\mathrm{CM}(S) = k$.*

Proof. As the largest jar has more cookies than all the other jars together, any strategy has to include a step in which Cookie Monster takes cookies from the largest jar only. Cookie Monster will not jeopardize the strategy if he takes all the cookies from the largest jar on the first move. Applying the induction process, we see that we need at least n moves. □

The lexicographically first sequence that satisfies the lemma is the sequence of powers of two. The following lemma [4] shows that any sequence with growth smaller than powers of two can be emptied in fewer then k moves.

Lemma 3. *Let S be a set of k jars with $s_1 < s_2 < \ldots < s_k$ cookies. Then we have*

$$\mathrm{CM}(S) \leq \lfloor \log_2 s_k \rfloor + 1.$$

Proof. We use the binary algorithm. Choose m such that $2^{m-1} \leq s_k \leq 2^m - 1$. Let the set of move amounts be $A = \{2^0, 2^1, \ldots, 2^{m-1}\}$. Any integer less than 2^m can be expressed as a combination of entries in A by using its binary representation. Because all elements in S are less than 2^m, all s_i may be written as sums of subsets of A. In other words, each jar in S may be emptied by a sum of move amounts in A. Therefore, $CM(S) \leq |A| = m$. However, $2^{m-1} \leq s_k$ implies that $m \leq 1 + \log_2 s_k$. Therefore, $CM(S) \leq m \leq 1 + \lfloor \log_2 s_k \rfloor$. $\qquad\square$

It follows that the lower bound in Theorem 1 is achieved for sets of k jars with the largest jar less than 2^m. The lexicographically smallest such sequence is the sequence of natural numbers: $S = 1, 2, 3, 4, \ldots, k$.

We note that if we multiply the number of cookies in every jar by the same number, the Cookie Monster number of that set of jars does not change. It follows that the lower bound can also be reached for sequences where all s_i are divisible by d and $s_k < d \cdot 2^m$.

Cavers [4] presented a corollary of Lemma 3 that bounds $CM(S)$ by the difference between the maximum and minimum elements of the set S.

Corollary 1. *Let S be a set of k jars with $s_1 < s_2 < \ldots < s_k$ cookies. Then it follows that*

$$CM(S) \leq 2 + \lfloor \log_2(s_k - s_1) \rfloor.$$

Proof. We remove s_1 cookies from every jar on the first move. $\qquad\square$

It follows that if S forms an arithmetic progression, then $CM(S) \leq 2 + \lfloor \log_2(k - 1) \rfloor$.

4 Naccis

We now present our Cookie Monster with interesting sequences of cookies in his jars. First, we challenge our monster to empty a set of jars containing cookies in the Fibonacci sequence. The Fibonacci sequence is defined as $F_0 = 0$, $F_1 = 1$, and $F_i = F_{i-2} + F_{i-1}$ for $i \geq 2$. A jar with zero cookies and one of the two jars containing one cookie are irrelevant, so our smallest jar will contain F_2 cookies. Belzner [1] found that the set S of k jars containing $\{F_2, F_3, \ldots, F_{k+1}\}$ cookies has $CM(S) = \lfloor \frac{k}{2} \rfloor + 1$. To present our proof, we need the following well-known equation [5] for the Fibonacci numbers.

Lemma 4. *For the set of Fibonacci numbers $\{F_1, F_2, \ldots, F_{k+1}\}$, we have*

$$F_{k+1} - 1 = \sum_{i=1}^{k-1} F_i.$$

More precisely, we need the following corollary.

Corollary 2. *The Fibonacci numbers satisfy*

$$F_{k+1} > \sum_{i=1}^{k-1} F_i.$$

Theorem 2. *For k jars with the set of cookies* $S = \{F_2, \ldots, F_{k+1}\}$, *the Cookie Monster number is* $\mathrm{CM}(S) = \lfloor \frac{k}{2} \rfloor + 1$.

Proof. By Corollary 2, one of Cookie Monster's moves when emptying the set $S = \{F_2, \ldots, F_{k+1}\}$ must touch the kth jar containing F_{k+1} cookies and not the first $k - 2$ jars containing the set of $\{F_2, \ldots, F_{k-1}\}$ cookies. Cookie Monster cannot be less optimal if he starts with this move. Suppose that on his first move, the monster takes F_k cookies from the $(k - 1)$st and kth jars. This empties the $(k - 1)$st jar and reduces the kth jar to $F_{k+1} - F_k = F_{k-1}$ cookies, which is the same number of cookies that the $(k - 2)$nd jar has. Thus, two jars are discarded in one move. He cannot do better than that. By induction, Cookie Monster may optimally continue in this way. If k is odd, the monster needs an extra move to empty the last jar, and the total number of moves is $(k + 1)/2$. If k is even, Cookie Monster needs two extra moves to empty the last two jars, and the total number of moves is $k/2 + 1$. Therefore, for any k, we have

$$\mathrm{CM}(S) = \left\lceil \frac{k+1}{2} \right\rceil = \left\lfloor \frac{k}{2} \right\rfloor + 1.$$

$\square$

There exist lesser-known and perhaps more challenging sequences of numbers similar to Fibonacci called *n-naccis* [6]. We define the *n*-nacci sequence as $N_i = 0$ for $0 \leq i < n - 1$, $N_i = 1$ for $n - 1 \leq i < n$, and $N_i = N_{i-n} + N_{i-n+1} + \cdots + N_{i-1}$ for $i \geq n$. For example, in the 3-nacci sequence, otherwise known as Tribonacci, each term after the third is the sum of the previous three terms. The Tetranacci sequence is made by summing the previous four terms, the Pentanacci by the previous five, and the construction of higher *n*-naccis follows.

We start with the Tribonacci sequence, denoted T_i: 0, 0, 1, 1, 2, 4, 7, 13, 24, 44, 81, ... The main property of the Tribonacci sequence, like the Fibonacci sequence, is that the next term is the sum of previous terms. We can use this fact to produce a similar strategy for emptying jars with Tribonacci numbers. We prove that the set S of k jars containing $S = \{T_3, \ldots, T_{k+2}\}$ cookies has $\mathrm{CM}(S) = \lfloor \frac{2k}{3} \rfloor + 1$ by using the inequalities relating the Tribonacci numbers proved in Lemma 5.

Lemma 5. *The Tribonacci sequence satisfies the inequalities*

- $T_{k+1} > \sum\limits_{i=1}^{k-1} T_i$, *and*
- $T_{k+2} - T_{k+1} > \sum\limits_{i=1}^{k-1} T_i$.

Proof. We proceed by induction. For the base case $k = 1$, both inequalities are true. Suppose they are both true for some k. If T_k is added to both sides of both inequalities' induction hypotheses, we obtain

$$T_{k+1} + T_k > \sum_{i=1}^{k} T_i,$$

and

$$T_{k+2} - T_{k+1} + T_k > \sum_{i=1}^{k} T_i.$$

But it is true that $T_{k+1} + T_k = T_{k+3} - T_{k+2}$ and $T_{k+2} - T_{k+1} + T_k < T_{k+2}$. Therefore, we have

$$T_{k+3} - T_{k+2} = T_{k+1} + T_k > \sum_{i=1}^{k} T_i,$$

$$T_{k+2} > T_{k+2} - T_{k+1} + T_k > \sum_{i=1}^{k} T_i;$$

both inequalities are true for $k + 1$. □

Theorem 3. *When k jars contain a set of Tribonacci numbers $S = \{T_3, \ldots, T_{k+2}\}$, then we have* $CM(S) = \lfloor \frac{2k}{3} \rfloor + 1$.

Proof. Consider the largest three jars. The largest jar and the second-largest jar each have more cookies than the remaining $k - 3$ jars do in total. That means Cookie Monster must perform a move that includes the second-largest, and possibly the largest, jars and does not touch the smallest $k - 3$ jars. Because of the second inequality in Lemma 5, the largest jar needs to be reduced by one more move that does not touch the smallest $k - 3$ jars to be discarded. Thus, because at least two moves are needed to touch and discard the three largest jars, discarding all three jars in two moves is optimal.

This can be achieved if Cookie Monster takes T_{k+1} cookies from the two largest jars on his first move, emptying the second-largest jar and reducing the number of cookies in the largest jar to $T_{k+2} - T_{k+1}$. He then should take T_k cookies from the third-largest and the largest jars on his second move, emptying the third-largest jar and reducing the largest jar to $T_{k+2} - T_{k+1} - T_k = T_{k-1}$ cookies. Now the largest jar is discarded, as the number of cookies left there is the same as the number of cookies in another jar. Since this strategy empties three jars in two moves, the monster may optimally continue in this way.

If k has remainder 1 modulo 3, he needs one more move for the last jar, bringing the total to $2\lfloor k/3 \rfloor + 1$. If k has remainder 2 modulo 3, he needs two more moves, and the total is $2\lfloor k/3 \rfloor + 2$. If k has remainder 0 modulo 3, he needs three moves for the last group of three, bringing the total to $2\lfloor k/3 \rfloor + 1$. Therefore, for any k, we have

$$\mathrm{CM}(S) = \left\lfloor \frac{2k}{3} \right\rfloor + 1.$$

$\square$

More generally, here is Cookie Monster's strategy for dealing with n-nacci sequences, which we call *cookie-monster-knows-addition*. He takes $n-1$ moves to empty the n largest jars. On the ith move, he takes N_{k-i} cookies from the $(k - i)$th-largest jar and the largest jar. This way he empties the $(k - i)$th-largest jar and reduces the kth-largest jar for each i such that $0 < i < n$. In doing this, n jars are emptied in $n - 1$ moves. This process can be repeated, until at most n elements remain, which he empties one by one. Thus, when $S = \{N_n, \ldots, N_{n+k-1}\}$, we will prove that the Cookie Monster number is

$$\mathrm{CM}(S) = \left\lfloor \frac{(n - 1)k}{n} \right\rfloor + 1.$$

We first prove the necessary inequalities relating the n-nacci numbers.

Lemma 6. *The n-nacci sequence satisfies the inequality*

$$N_{k+1} > \sum_{i=1}^{k-1} N_i.$$

Proof. The proof is the same as in the well-known equality for the Fibonacci numbers [5] (see Lemma 4) and Corollary 2. $\square$

This inequality is not surprising when we notice that the a-nacci sequence grows faster than the b-nacci sequence for $a > b$. Hence, other n-nacci sequences grow faster than the Fibonacci sequence. But the following inequality is more subtle.

Lemma 7. *The n-nacci sequence satisfies the inequality*

$$N_{k+n-1} - \sum_{i=k+1}^{k+n-2} N_i > \sum_{i=1}^{k-1} N_i.$$

Proof. By the definition, we have $N_{k+n-1} - \sum_{i=k+1}^{k+n-2} N_i = N_k + N_{k-1}$. By Lemma 6, it follows that $N_k > \sum_{i=1}^{k-2} N_i$. Hence, we have $N_k + N_{k-1} > \sum_{i=1}^{k-1} N_i$. □

The next theorem shows that *n*-nacci numbers satisfy many inequalities between the two presented above.

Theorem 4. *The n-nacci sequence satisfies the following inequality, for any* $0 \le j \le n - 2$:

$$N_{k+j} - \sum_{i=k+1}^{k+j-1} N_i > \sum_{i=1}^{k-1} N_i.$$

Proof. By definition, $N_{k+j} - \sum_{i=k+1}^{k+j-1} N_i = \sum_{i=k+j-n}^{k} N_i$. By the inequality in Lemma 6, we have

$$\sum_{i=k+j-n}^{k} N_i = N_k + \sum_{i=k+j-n}^{k-1} N_i > \sum_{i=1}^{k-2} N_i + N_{k-1} = \sum_{i=1}^{k-1} N_i.$$

□

Now we are equipped to come back to the Cookie Monster number.

Theorem 5. *When k jars contain a set of n-nacci numbers* $S = \{N_n, \ldots, N_{n+k-1}\}$, *the Cookie Monster number is*

$$CM(S) = \left\lfloor \frac{(n-1)k}{n} \right\rfloor + 1.$$

Proof. Consider the largest *n* jars. The largest $n-1$ jars each have more cookies than the remaining $k - n$ jars do in total. That means the Cookie Monster must perform a move that includes the largest $n - 1$ jars and does not touch the smallest $k - n$ jars. Suppose he touches the $(n - 1)$st-largest jar on his first move. After that, even if he took cookies from the largest $n - 2$ jars during his previous move, the $(n - 2)$nd-largest jar will still have more cookies than all of the smallest $k - n$ jars combined (due to the inequalities in Theorem 4). Thus there must be another move that touches the $(n-2)$nd-largest jar and does not touch the smallest $k - n$ jars nor touches the $(n-1)$st-largest jar. Continuing this line of reasoning, there should be a move that touches the $(n-3)$rd-largest jar and does not touch the smallest $k - n$ jars, nor does it touch the $(n-1)$st-largest jar nor the $(n-2)$nd-largest jar, and so on.

Summing up for every jar among the $n - 1$ largest jars, there is a move that touches it and possibly the jars larger than it as well as the nth-largest jar, but does not touch anything else. Hence, there must be at least $n - 1$ moves that do not touch the smallest $k - n$ jars.

We know that we can empty the largest n jars in $n - 1$ moves if Cookie Monster uses his *cookie-monster-knows-addition* strategy. Thus, because at least $n - 1$ moves are needed to touch and discard the last n jars, discarding all n jars in $n - 1$ moves is optimal.

We can continue doing this until we have no more than n jars left. Because the smallest n jars in set S are powers of two, we must use the same number of moves as the number of jars left to empty the remaining jars. If k has nonzero remainder x modulo n, then Cookie Monster needs x additional moves for the last jars. Hence, the total number of moves is $(n - 1)\lfloor k/n \rfloor + x$. If k has remainder 0 modulo n, he needs n additional moves to empty the final n jars for a total of $(n - 1)\lfloor k/n \rfloor + 1$ moves. Therefore, for any k, we save one move for every group of n jars besides the last n jars. Hence, we save $\lfloor \frac{k-1}{n} \rfloor$ moves, and the Cookie Monster number of S is

$$\mathrm{CM}(S) = k - \left\lfloor \frac{k-1}{n} \right\rfloor = \left\lfloor \frac{(n-1)k}{n} \right\rfloor + 1. \qquad \square$$

5 Super Naccis

The monster wonders whether he can extend his knowledge of nacci sequences to non-nacci ones. He first considers Super-n-nacci sequences that grow at least as fast as n-nacci sequences. Define a *Super-n-nacci sequence* as $S = \{M_1, \ldots, M_k\}$, where $M_i \geq M_{i-n} + M_{i-n+1} + \cdots + M_{i-1}$ for $i \geq n$. Cookie Monster suspects that since he already knows how to consume the nacci sequences, he might be able to bound $\mathrm{CM}(S)$ for Super naccis.

Theorem 6. *For Super-n-nacci sequences S with k terms, we have*

$$\left\lfloor \frac{(n-1)k}{n} \right\rfloor + 1 \leq \mathrm{CM}(S).$$

Proof. The proof of the bound for n-nacci sequences (see Theorem 4) used only the inequalities. The same proof works here. $\qquad \square$

6 Beyond Naccis

We found sequences representing k jars such that their Cookie Monster numbers are asymptotically rk, where r is a rational number of the form $(n-1)/n$. Is it possible to invent other sequences whose Cookie Monster numbers are asymptotically rk, where r is any rational number not exceeding 1?

Before discussing sequences and their asymptotic behavior, we go back to the bounds on the Cookie Monster number of a set and check to see whether any value between the bounds is achieved.

First, we need a definition. A set $S = \{s_1, s_2, \ldots, s_k\}$ of increasing numbers s_i is called *two-powerful* if it contains all the powers of 2 not exceeding $\max(S) = s_k$. We can calculate the Cookie Monster number of a two-powerful set.

Lemma 8. *Given a two-powerful set* $S = \{s_1, s_2, \ldots, s_k\}$, *its Cookie Monster number is the smallest m such that 2^m is larger than all elements in S:* $\mathrm{CM}(S) = \lfloor \log_2 s_k \rfloor + 1$.

Proof. Let m be the smallest power of 2 not in S: $m = \lfloor \log_2 s_k \rfloor + 1$. Then S contains a subset of powers of 2, namely $S' = \{2^0, 2^1, \ldots, 2^{m-1}\}$. This subset has a Cookie Monster number m. A superset of S' cannot have a smaller Cookie Monster number, so $\mathrm{CM}(S) \geq m$. But by Lemma 3, we see that $\mathrm{CM}(S) \leq m$. Hence we have $\mathrm{CM}(S) = m$. $\qquad\square$

Two-powerful sets are important because they are easy to construct, and we know their Cookie Monster numbers. They become crucial building blocks in the following theorem.

Theorem 7. *For any k and m such that $m \leq k < 2^m$, there exist a set S of jars of length k such that* $\mathrm{CM}(S) = m$.

Proof. The given constraint allows us to build a two-powerful set S of length k such that $2^{m-1} \leq s_k < 2^m$. We include in this set all powers of 2 from 1 to 2^{m-1} and any other $k - m$ numbers not exceeding 2^m. This two-powerful set satisfies the condition. $\qquad\square$

Now we return to sequences. Suppose $s_1, s_2, \ldots$ is an infinite increasing sequence. Let us denote the set of the first k elements of this sequence as S_k. We are interested in the ratio of $\mathrm{CM}(S_k)/k$ and its asymptotic behavior.

If we have $s_i = 2^{i-1}$, then $\mathrm{CM}(S_k)/k = 1$. If $s_i = i$, then it follows that $\mathrm{CM}(S_k)/k = (\lfloor \log_2 k \rfloor + 1)/k$, which tends to zero when i tends to infinity. We know that for Fibonacci numbers the ratio is $1/2$; for Tribonacci, it is $2/3$; and for n-naccis, it is $(n-1)/n$. What about other ratios? Are they possible?

Yes. We claim that any ratio $r : 0 \leq r \leq 1$ is possible. We prove this by constructing sequences with any given r. The idea is to take a sequence that contains all the powers of 2 and to add some numbers to the sequence as needed. Let us first construct the sequence explicitly.

6.1 The Sequence

We build the sequence by induction. We start with $s_1 = 1$. Then it follows that $CM(S_1)/1 = 1 \geq r$. We process natural numbers one by one and decide whether to add a number to the sequence by the following rules:

- If it is a power of 2, we always add it.
- If it is not a power of 2, we add it if the resulting ratio does not go below r.

Now we would like to study the sequence and prove some lemmas regarding it. Let us denote the elements of this sequence by s_i, its set of first k elements by $S_k = \{s_1, s_2, \ldots, s_k\}$, and the ratio $CM(S_k)/k$ by r_k. By the construction we have, $r_k \geq r$. We need to prove that $\lim_{k \to \infty} r_k = r$.

Suppose $CM(S_k) = m$, so that the current ratio r_k is m/k. If s_{k+1} is a power of 2, then $r_{k+1} = (m + 1)/(k + 1)$ and the difference is

$$r_{k+1} - r_k = \frac{k - m}{k(k + 1)}.$$

As $0 \leq k - m < k$, we get

$$0 \leq r_{k+1} - r_k \leq \frac{1}{k + 1}.$$

In this case the ratio does not decrease, but the increases are guaranteed to be smaller and smaller as k grows. If s_{k+1} is not a power of 2, then $r_{k+1} = m/(k + 1)$ and the difference is $r_{k+1} - r_k = -m/k(k + 1)$. In this case the ratio always decreases.

Lemma 9. *If $r = 1$, then the sequence contains only powers of 2. If $r = 0$, then the sequence contains all the natural numbers.*

Proof. We start with the ratio 1 for the first term of the sequence: $r_1 = 1$. Every nonpower of 2 in the sequence decreases the ratio. So if $r = 1$, we cannot include nonpowers of 2. If $r = 0$, the ratio r_k is always positive, so we include every nonpower of 2. □

The sequences in the previous lemma produce the ratios 0 and 1, so from now on we can assume that $0 < r < 1$. Let us see what happens if we include all numbers between two consecutive powers of 2 in the sequence. Since all powers of 2 are present in the sequence, let us denote the index of 2^m in the sequence by k_m: $s_{k_m} = 2^m$. Hence, $CM(S_k) = m$ if $k_{m-1} \le k < k_m$. Also, we have $r_{k_m} = (m+1)/k_m$.

Lemma 10. *If we include all the nonpowers of 2 in the sequence between k_m and k_{m+1}, then the ratio of ratios is bounded: $r_{k_{m+1}}/r_{k_m} \le (m+2)/2(m+1)$.*

Proof. Suppose by the algorithm we need to add all the numbers between k_m and k_{m+1} to the sequence. Therefore we have $k_{m+1} = k_m + 2^m$. The ratios are then $r_{k_m} = (m+1)/k_m$ and $r_{k_{m+1}} = (m+2)/(k_m + 2^m)$. So the ratio of ratios is

$$\frac{r_{k_{m+1}}}{r_{k_m}} = \frac{m+2}{m+1} \cdot \frac{k_m}{k_m + 2^m}.$$

Using the fact that $k_m \le 2^m$, we get $r_{k_{m+1}}/r_{k_m} \le (m+2)/2(m+1)$. So as m grows, the ratio is almost halved. Starting from $m = 3$, we can guarantee that this ratio is never more than $2/3$. □

Corollary 3. *If we include all the nonpowers of 2 for $m > 2$, then the ratio of ratios is bounded by a constant that does not depend on m: $r_{k_{m+1}}/r_{k_m} < 2/3$.*

6.2 The Theorem

Now we are ready to prove the following theorem.

Theorem 8. *For any real number r with $0 \le r \le 1$, there exists a sequence s_i with sets of first k elements $S_k = \{s_1, s_2, \ldots, s_k\}$ that have Cookie Monster numbers such that $CM(S_k)/k$ tends to r when k tends to infinity.*

Proof. As we mentioned before, we can assume that $0 < r < 1$. Given r, we build the sequence described in Section 6.1. While building the sequence, if we need to skip the next number, we have approached r within $m/k(k+1)$:

$$r \le r_k \le r + \frac{m}{k(k+1)} \le r + \frac{1}{k+1}.$$

If our sequence contains all but a finite number of natural numbers, r_k tends to zero. Since the ratio should never go below r, we get a contradiction. Hence, we must drop infinitely many numbers. Each time we drop a number, the ratio r_k gets within $1/(k+1)$ of r. Therefore, with each next number dropped, we get closer and closer to r. Now we must prove that not only can we get as close to r as we want, but also that we do not wander off too far from it in between.

Take ϵ such that $\epsilon < r/6$, and consider k such that $1/(1+k) < \epsilon$. We can find a number t such that $t > k$ and $r_t < r + \epsilon$. Thus we have approached r within the distance of ϵ, and we continue building the sequence. If the next number is a nonpower of 2, then the ratio approaches r even closer. When we reach the next power of 2, then the ratio increases by no more than ϵ. Therefore the ratio stays within 2ϵ of r, so it will not exceed $4r/3$.

We claim that after this power of 2, we cannot add all nonpowers of 2 until the next power of 2. Indeed, if that were the case, then the ratio would drop to a number below $4r/3 \cdot 2/3 < r$. Therefore we have to drop a nonpower of 2 from the sequence after the first encountered power of 2. We then approach the ratio again and get at least ϵ-close to it. Thus, for numbers greater than t, the ratio will never be more than 2ϵ away from r. $\square$

Acknowledgments

Cookie Monster is proud that people study his cookie-eating strategy. Cookie Monster and the authors are grateful to the MIT-PRIMES program for supporting this research.

References

[1] M. Belzner. Emptying sets: The Cookie Monster problem. math.CO arXiv:1304.7508 (accessed August 2013).

[2] O. Bernardi and T. Khovanova. The Cookie Monster problem. http://blog. tanyakhovanova.com/?p=325 (accessed August 2013).

[3] P. Vaderlind, R. K. Guy, and L. C. Larson. *The Inquisitive Problem Solver*. Mathematical Association of America, Washington, DC, 2002.

[4] M. Cavers. Cookie Monster problem notes. University of Calgary Discrete Math Seminar, private communication, 2011.

[5] Wikipedia: The Free Encyclopedia. Fibonacci numbers. http://en.wikipedia.org/ w/index.php?title=Fibonacci_number&oldid=572271013 (accessed August 2013).

[6] Wikipedia: The Free Encyclopedia. Generalizations of Fibonacci numbers. http://en.wikipedia.org/w/index.php?title=Generalizations_of_Fibonacci_numbers &oldid=560561954 (accessed August 2013).

17

◇◇

REPRESENTING NUMBERS USING
FIBONACCI VARIANTS

Stephen K. Lucas

The Fibonacci numbers and their variants are among the most popular sequences in recreational mathematics and even have a journal (*The Fibonacci Quarterly*) dedicated to them. Many books have been written about them, including Dunlap [4] and Vajda [17]. Fibonacci numbers are defined by the recurrence relation $f_n = f_{n-1} + f_{n-2}$ with initial conditions $f_0 = 0$ and $f_1 = 1$. They have the closed form representation $f_k = \left(\phi^k - (1 - \phi)^k\right)/\sqrt{5}$, where $\phi = (1 + \sqrt{5})/2$ is the well-known golden ratio, also with its own book, by Livio [13].

One property of Fibonacci numbers is that every natural number can be uniquely represented by a sum of distinct nonconsecutive Fibonacci numbers starting from $f_2 = 1$, since $f_1 = f_2 = 1$ is a repeat, and $f_0 = 0$ won't contribute to a sum. This was first discovered by Eduourd Zeckendorf in 1939, but only published by him in 1972 [19]. The first publication of the result was by Lekkerkerker [12] in 1952. As a result, representing a natural number in this form is known as its Zeckendorf representation, and is easily found using a greedy algorithm: given a number, subtract the largest Fibonacci number less than or equal to it, and repeat until the entire number is used up. For example, consider 825. The largest Fibonacci number less than 825 is $f_{15} = 610$, and $825 - 610 = 215$. Then the procedure gives $f_{12} = 144$ and $215 - 144 = 71$, $f_{10} = 55$ and $71 - 55 = 16$, $f_7 = 13$ and $16 - 13 = 3$, and finally $f_4 = 3$. Therefore, the number is given by $825 = f_{15} + f_{12} + f_{10} + f_7 + f_4$, which can be represented as $(10010100100100)_Z$, where the Z indicates Zeckendorf representation, and digits from right to left indicate whether or not f_2, f_3, and so on are included in the sum for the number.

The lack of consecutive Fibonacci numbers in Zeckendorf representations of natural numbers means a pair of ones can be used to separate numbers in a list. This means different numbers in a list can be represented using a different number of digits, known as a *variable-length encoding*. Traditionally, a fixed number of digits (in a given base) is used to represent every number in a list, which may waste space. The first part of this chapter compares the efficiency of representing numbers using Zeckendorf form vs. traditional binary with a fixed number of digits and shows when Zeckendorf form is to be preferred. We shall also see what happens when variants of Zeckendorf form are used.

Not only can we represent natural numbers as sums of Fibonacci numbers, we can also do arithmetic with them directly in Zeckendorf form. The chapter includes a survey of past approaches to Zeckendorf representation arithmetic, as well as some improvements.

1 Zeckendorf Proofs

The proof that Zeckendorf representation exists and is unique for every natural number is straightforward enough to include here. Existence is proven by induction and begins with $1 = f_2, 2 = f_3$, and $3 = f_4$. If we assume every natural number up to n has a Zeckendorf representation, consider $n + 1$. If it is a Fibonacci number, we are done. Otherwise, there is some j such that $f_j < n + 1 < f_{j+1}$. Since $n + 1 - f_j < n$, it has a Zeckendorf representation, and additionally, since $n + 1 - f_j < f_{j+1} - f_j = f_{j-1}$, it doesn't contain f_j or f_{j-1}. Thus the Zeckendorf representation of $n + 1$ is that for $n + 1 - f_j$ with f_j included, and each Fibonacci number in the representation occurs at most once and nonconsecutively. By induction, we are done.

Before proving uniqueness, we need the result that any sum of distinct nonconsecutive Fibonacci numbers whose largest is f_n is strictly less than f_{n+1}. Again, we can use induction, and since $f_2 = 1 < f_3 = 2$, $f_3 = 2 < f_4 = 3$ and $f_2 + f_4 = 4 < f_5 = 5$, it is initially true. Now assume that any sum of distinct nonconsecutive Fibonacci numbers whose largest is f_n is strictly less than f_{n+1}. Then a sum with largest number f_{n+1} can be split into f_{n+1} and a sum with largest Fibonacci number at most f_{n-1}, which is strictly less than f_n by the assumption. The combination is thus strictly less than $f_n + f_{n+1} = f_{n+2}$, and we are done.

And now to uniqueness. Assume that there are two different sets of nonconsecutive Fibonacci numbers that have the same sum, A and B. If there are any Fibonacci numbers common to both collections, remove them from both by set difference, and let C and D be the differences $C = A - B$ and $D = B - A$. Since the same Fibonacci numbers are being removed from both A and B, the Fibonacci numbers in the smaller sets C and D still have the same sum, and have no common numbers. Let f_c and f_d be the largest elements of C and D, respectively, and since there are no common numbers, $f_c \neq f_d$. In addition, without loss of generality, assume $f_c > f_d$. But by our previous result, the sum of D is strictly less than f_{d+1} and so must also be strictly less than f_c. But the sum of C is at least f_c. The only way this is possible is for C and D to in fact be empty sets that sum to zero. But this means that the sets A and B must be the same, and we do have a unique representation.

2 Efficiency of Number Representations

A number's Zeckendorf representation, being a string of zeros and ones, looks a lot like a base two representation. They are, however, quite different. For example, the base two representation of 825, which equals $512 + 256 + 32 + 16 + 8 + 1$, is $(1100111001)_2$. This requires fewer digits than its Zeckendorf representation and so is more efficient with regard to the number of digits required. In fact, the base two representation of a natural number will always be shorter than its Zeckendorf representation, because the larger the base, the smaller the number of digits required. Zeckendorf form does not formally have a base, but since $\phi \approx 1.618$ and $1 - \phi \approx -0.618$, f_k is the closest natural number to $\phi^k/\sqrt{5}$, so the ratio between Fibonacci numbers approaches ϕ. Thus Zeckendorf representation roughly has the base $\phi \approx 1.618 < 2$, and so is less efficient than binary.

2.1 Lists of Natural Numbers

While Zeckendorf representation is less efficient (with regard to the number of digits required) than binary, its big advantage is that it cannot contain a pair of consecutive ones. So instead of using a fixed number of digits to represent individual natural numbers in a list, as many digits as are necessary can be used for each number, with a pair of ones separating consecutive numbers. We can do even better if we reverse the order of the digits, with least-to most-significant digits from left to right. Using the reverse of the standard way we write numbers, the rightmost digit of a number's Zeckendorf representation will always be a one. So when a pair of ones is used to separate numbers, the first is part of the first number, and only the second is the separator between numbers. For example, the stream of digits "10010101110001011011" can be separated out into "10010101," "1000101," and "01," or $f_2 + f_5 + f_7 + f_9$, $f_2 + f_6 + f_8$, and f_3, or 53, 30, and 2, respectively.

This reversed way of representing numbers in a list is sometimes known as *Fibonacci coding* and is particularly useful when there is no prior knowledge of the range of the list of numbers. In this case, using a base two representation with a fixed number of binary digits (bits) is problematic. We might overestimate how big the numbers will get and waste space in base two, or underestimate and have numbers we can't represent.

Even when we know exactly what range of numbers to expect in a list, there are occasions when Fibonacci coding is superior to traditional base two. For example, consider a list of numbers known to range from one to a million, with each number being equally likely. A base two representation of numbers in this range will require at least twenty bits per number. Using Fibonacci coding, the same distribution of numbers requires on average 27.8 bits per number, so it uses more space and thus is less efficient. But if the numbers from one to ten are equally likely (with probability 9,999/100,000) and one million occurs with probability only one in ten thousand, then the base two list stays at twenty bits per number, but the Fibonacci coding reduces to only 4.6 bits per number on average. While this may be an extreme case, the saving is substantial, particularly when smaller numbers are more likely. As another example, consider nonnegative integers chosen with Poisson distribution ($P(X = k) = \lambda^k e^{-\lambda}/k!$) and $\lambda = 4$. Practically, output is integers from one to thirty-one, and Fibonacci coding requires 4.6 bits per number. Binary would require five bits per number. In these last two cases, lists of numbers are more efficiently written using Fibonacci coding.

2.2 Arbitrary Reals and Continued Fractions

There is one area where Fibonacci coding has not previously been applied and is particularly appropriate—the representation of arbitrary reals as continued fractions. A continued fraction representation of a real is essentially a list of natural numbers. The list has no a priori upper bound, and smaller natural numbers occur more often than larger ones.

As a reminder, continued fractions are closely related to the algorithm for finding the greatest common divisor of two natural numbers using integer division. For example, given 236 and 24, we can successively find $236 = 9 \times 24 + 20$, then $24 = 1 \times 20 + 4$, and $20 = 5 \times 4 + 0$, which tells us that gcd(236, 24) = 4. But we can also use these steps to write

$$\frac{236}{24} = 9 + \frac{20}{24} = 9 + \frac{1}{24/20} = 9 + \frac{1}{1 + \frac{4}{20}} = 9 + \frac{1}{1 + \frac{1}{20/4}} = 9 + \frac{1}{1 + \frac{1}{5}}.$$

Any fraction can be rewritten in this form of fractions within fractions (hence the name "continued fractions"), for which all the fractions have 1 as their numerator. A common, more compact, notation rewrites this as $236/24 = [9; 1, 5]$. In general, a simple continued fraction for a (positive) fraction is

$$\frac{p}{q} = b_0 + \cfrac{1}{b_1 + \cfrac{1}{b_2 + \cfrac{\vdots}{b_{n-1} + \cfrac{1}{b_n}}}} \equiv [b_0; b_1, b_2, \ldots, b_n],$$

where b_0 is an integer, and the b_is, for $i > 0$, are natural numbers. The compact notation on the right is much more convenient when the number of terms in the continued fraction representation becomes large. The b_is are traditionally called "partial quotients." Let us use the notation $\lfloor x \rfloor$, stated "the floor of x," for the largest integer less than or equal to x. Then the algorithm for finding partial quotients (adapted from integer division) is: given some real number x, set $x_0 = x$ and $b_0 = \lfloor x_0 \rfloor$, then we have

$$x_i = \frac{1}{x_{i-1} - b_{i-1}} \quad \text{and} \quad b_i = \lfloor x_i \rfloor \quad \text{for} \quad i = 1, 2, \dots.$$

If x is rational, eventually some x_i will be an integer, and the continued fraction terminates. If x is irrational, the sequence of partial quotients goes forever. For our example with $x = 236/24$, the result is

$$x_0 = \frac{236}{24}, \qquad\qquad b_0 = \left\lfloor \frac{236}{24} \right\rfloor = 9,$$

$$x_1 = \frac{1}{236/24 - 9} = \frac{1}{5/6} = 6/5, \qquad b_1 = \left\lfloor \frac{6}{5} \right\rfloor = 1,$$

$$x_2 = \frac{1}{6/5 - 1} = \frac{1}{1/5} = 5, \qquad b_3 = \lfloor 5 \rfloor = 5,$$

and we are done.

Continued fractions have many elegant features, and a straightforward introduction is Olds [15]. Many introductory texts on number theory, including the classic Hardy and Wright [10], includes a chapter on continued fractions, but the feature of relevance here is the Gauss-Kuzmin theorem, which tells us that for almost all irrationals between zero and one, as $n \to \infty$ the probability that the nth partial quotient is k is

$$\lim_{n \to \infty} P(k_n = k) = -\log_2 \left(1 - \frac{1}{(k+1)^2} \right).$$

Khinchin [11] gives an especially clear derivation of this result. Its importance from our perspective is that arbitrarily large partial quotients are possible, but increasingly unlikely, in a continued fraction representation. Table 17.1 shows the probabilities of the first few natural numbers occurring as any given partial quotient in the continued fraction representation of an arbitrary irrational, as well as the probabilities that large partial quotients can occur. Thus Fibonacci coding is an ideal choice for representing continued fraction partial quotients for arbitrary irrationals.

For example, consider ln(2), whose first twenty thousand partial quotients are summarized in Table 17.2. Since there are only twenty thousand partial quotients to work with, some of the probabilities aren't exactly equal to those in the theoretical distribution, but they are remarkably close. The largest partial quotient happens to be 963,664. If we knew this beforehand, we could encode this continued fraction using binary with twenty bits per number. The

TABLE 17.1.
Probabilities of various partial quotients occurring in a random continued fraction

k	Probability	k	Probability
1	0.415037	10	0.011973
2	0.169925	100	1.41434×10^{-4}
3	0.093109	1000	1.43981×10^{-6}
4	0.058894	10,000	1.44241×10^{-8}
5	0.040642		
6	0.029747	> 10	1.25531×10^{-1}
7	0.022720	> 100	1.42139×10^{-2}
8	0.017922	> 1000	1.44053×10^{-3}
9	0.014500	>10,000	1.44248×10^{-4}

TABLE 17.2.
Probabilities of the first twenty thousand partial quotients occurring in ln(2)

k	Probability	k	Probability
1	0.4152	10	0.01285
2	0.1668	100	2.5×10^{-4}
3	0.09405	1000	0
4	0.0577	10,000	0
5	0.0397		
6	0.03065	> 10	1.284×10^{-1}
7	0.0222	> 100	1.45×10^{-2}
8	0.0179	> 1000	1.65×10^{-3}
9	0.0145	>10,000	1.5×10^{-4}

Fibonacci coding of these partial quotients requires on average the shockingly low 3.74 bits per number. Not only do we not need to know the largest number in the sequence beforehand, but we have an extraordinarily compact way of representing the sequence. For comparison, a version of Lochs' theorem [14, 18] states that if m is the number of terms in a number's continued fraction expansion and p is the number of correct digits in the continued fraction when converted to binary representation, then almost always the following holds:

$$\lim_{n \to \infty} \frac{m}{n} = \frac{6(\ln 2)^2}{\pi^2} \approx 0.292.$$

Assuming a long enough continued fraction, this tells us that a number with m partial quotients will be accurate to about $3.42m$ bits in base two. Since $3.74 > 3.42$, the binary representation of ln(2) is slightly more efficient than the Fibonacci coding of its continued fraction to about sixty-eight thousand binary digits, or about twenty thousand decimal digits. But we lose all the additional information given by the continued fraction partial quotients.

As another example, consider the first twenty thousand partial quotients of π. In this case, the largest partial quotient is 74,174, and the Fibonacci coding requires 3.71 bits per number. Slightly better than in the $\ln(2)$ case, but still slightly worse than the pure binary representation.

In conclusion, lists of natural numbers where smaller numbers occur more often than larger ones are more compactly represented using Fibonacci coding instead of traditional binary representations. This is particularly striking when representing lists of partial quotients for arbitrary irrationals, where in addition to varying magnitude, there is no prescribed upper bound. Unfortunately, the binary representation is still slightly more efficient than the Fibonacci coded continued fraction representation, if all you are interested in is a high-precision representation.

3 Generalizing Fibonacci Coding

We have already seen how the effective base of Zeckendorf representation is $\phi \approx 1.618 < 2$, so more digits are required than in traditional binary. However, there is no reason we have to be limited to traditional Fibonacci numbers. The *Fibonacci Quarterly* is filled with many variants on the Fibonacci sequence. The *tribonacci* sequence is defined by $t_n = t_{n-1} + t_{n-2} + t_{n-3}$, with $t_{-1} = t_0 = 0$ and $t_1 = 1$, and continues 1, 2, 4, 7, 13, 24, 44, ... The *tetranacci* sequence is defined by $u_n = u_{n-1} + u_{n-2} + u_{n-3} + u_{n-4}$ with $u_{-2} = u_{-1} = u_0 = 0$ and $u_1 = 1$, and continues 1, 2, 4, 8, 15, 29, 56, ... The earliest description of these series dates to 1963 in Feinberg [5]. As with Fibonacci numbers, every number can be uniquely represented as sums of tribonacci or tetranacci numbers. Three consecutive ones cannot appear in a tribonacci representation, and four consecutive ones cannot appear in a tetranacci representation. These are special cases of sequences of numbers that can be used to represent arbitrary integers, as described in Fraenkel [7]. Our interest in these representations is that the tribonacci numbers grow like x^n, where x is the largest root of $x^3 - x^2 - x - 1 = 0$, or about 1.8393; and tetranacci numbers grow like x^n, where x is the largest root of $x^4 - x^3 - x^2 - x - 1 = 0$, or about 1.9276. These roots are closer to two than the golden ratio, and so the number of digits required to represent natural numbers using these sequences will get closer to how many are required using binary digits, keeping the advantage that variable-length coding is possible. The disadvantage is that the number of repeated ones needed to separate numbers increases.

There is no reason we should stop with tetranacci numbers, although the naming conventions become cumbersome. Define a sequence of k-bonacci numbers by

$$u_n = \sum_{i=1}^{n} u_{n-i}, \quad \text{with } u_1 = 1 \text{ and } u_i = 0 \text{ for } i < 0.$$

TABLE 17.3.
Number of bits needed for k-bonacci representations

k	(a)	(b)	(c)	(d)	(e)
2	27.82	4.60	4.57	3.74	3.71
3	23.34	5.00	4.96	4.35	4.33
4	22.86	5.90	5.85	5.28	5.27
5	23.40	6.90	6.85	6.26	6.25

Notes: Shown are bits on average for k-bonacci representations for
(a) equally spaced, one to a million; (b) uneven distribution one to ten
and one million; (c) Poisson with $\lambda = 4$; (d) ln 2 partial quotients; and
(e) π partial quotients.

Then Fibonacci, tribonacci and tetranacci numbers are 2-bonacci, 3-bonacci, and 4-bonacci numbers, respectively. The 5-, 6-, and 7-bonacci numbers grow like 1.9659^n, 1.9836^n, and 1.9920^n, respectively, and the effective bases are approaching 2. Since $k - 1$ digits are used to separate different numbers in variable-length coding, there will be some optimum k-bonacci sequence to minimize the length of an encoding of a sequence of numbers, depending on their distribution.

Let us return to the examples where we compared Fibonacci coding to binary. Table 17.3 lists how many bits on average are required with k-bonacci coding for various values of k in a variety of examples. For uniformly distributed numbers, 4-bonacci numbers are the best sequence but are still inferior to binary when we happen to know the upper bound of a million. In every other case, Fibonacci coding is superior. Alas, the gains made by shortening the length of the sequence of digits required by increasing the effective base has been lost to the additional digits required to separate individual numbers.

4 Arithmetic

Not only can natural numbers be represented as sums of Fibonacci numbers, but also arithmetic can easily be done on them in this form. Here we return to Zeckendorf form with the most-significant digits on the left. After numbers are combined, the resulting sum of Fibonacci numbers will usually not be in Zeckendorf form, because it will include pairs of successive ones or numbers greater than one. Luckily, a sum of Fibonacci numbers can be easily returned to Zeckendorf form by a combination of two rules:

- The *pair rule*: since we have $f_n = f_{n-1} + f_{n-2}$ or $f_n - f_{n-1} - f_{n-2} = 0$, subtracting one from successive digits adds one to the digit immediately to their left, which can be represented by the transformation $(\ldots (+1)(-1)(-1) \ldots)_Z$.

- The *two rule*: subtracting $f_{n+1} = f_n + f_{n-1}$ from $f_n = f_{n-1} + f_{n-2}$, we get $f_{n+1} + f_{n-2} - 2f_n = 0$. Subtracting two from a digit adds one to the digit to its left (like an ordinary base two carry) and additionally adds one to the digit two to the right. This can be represented by the transformation $(\ldots(+1)(-2)(0)(+1)\ldots)$, where the (0) indicates no change to the digit. It is this nonstandard carry that makes arithmetic with Zeckendorf representation more entertaining.

The nonstandard carry with the two rule means we need to be more careful near the right edge of the number. Since $2f_1 = f_2$ and $2f_2 = f_3 + f_1$, the special cases of the two rule at the right edge are $(\ldots(+1)(-2))_Z$ and $(\ldots(+1)(-2)(+1))_Z$.

4.1 Addition

To add numbers in Zeckendorf form, we simply add the digits, then apply the pair and two rules as necessary to return to Zeckendorf form. For example, consider

$$(101001001)_F + (100101001)_F = (201102002)_F,$$

where the F indicates Fibonacci representation. The process of returning to Zeckendorf form is most easily visualized using a checkerboard representation, where boxes represent successive Fibonacci numbers and counters in a box indicate how many multiples of that Fibonacci number are needed. The pair and two rules govern how counters can be moved. Figure 17.1 shows one way of returning $(201102002)_F$ to Zeckendorf form. At the top we represent $(201102002)_F$. Then each successive row of the figure shows how applications of the pair rule (three times), the two rule (once), and the pair rule one more time leads to a distribution of counters in Zeckendorf form. The sum is thus $(10000010101)_Z$.

In this example, we did not apply a systematic approach when returning to Zeckendorf form. Early discussions of addition by Freitag and Phillips [8] in 1998 and Fenwick [6] in 2003 did not suggest a systematic approach. In 2002, Tee [16] suggested a recursive approach that had the rather pessimistic bound of $O(n^3)$, which states that the number of applications of pair or two rules required in returning to Zeckendorf form is proportional to the cube of the number of digits n. In 2013, Ahlbach et al. [1] showed how returning a sum to Zeckendorf form could be achieved using exactly $3n$ applications of pair or two rules. Their algorithm uses three passes through the digits, looking at groups of four successive digits. Specifically, their first pass from left to right performs the replacements (x is any digit) $020x \rightarrow 100(x + 1)$, $030x \rightarrow 110(x + 1)$, $021x \rightarrow 110x$, and $012x \rightarrow 101x$, which they show eliminates any twos in the

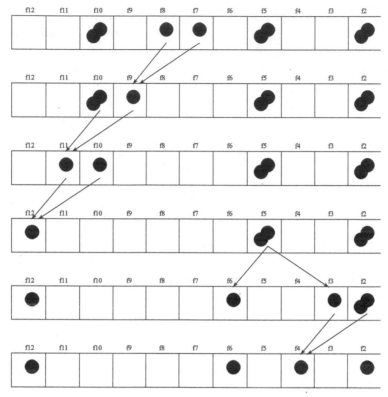

Figure 17.1. Using a checkerboard to visualize returning a number to Zeckendorf form.

representation combined with a clean-up operation at the end. Their second pass from right to left performs the replacement $011 \rightarrow 100$, which eliminates the pattern 1011 from the representation, and the third pass from left to right repeats the second pass and eliminates any remaining successive ones without additional consecutive ones.

Here, I introduce an improvement on Ahlbach et al. [1] that reduces the process to two passes. Start with the same first pass to remove twos. Then insert a leading zero, and the second pass from left to right does the replacement $(01)^k 1 \rightarrow 1(0)^{2k}$, where the notation $(01)^k$ means k copies of the pair of digits zero and one. This deals with any possible carries of long sequences of ones; it can be achieved by a single pass using a pair of pointers, one at the first 01 pair, the second moving to the right. If a pair of consecutive zeros are found, then the left pointer can be moved to match the right pointer, and we continue through the number. We cannot improve the efficiency further, due to the carry from the two rule going in both directions.

For example, let us reconsider $(201102002)_F$. Applying the first left-to-right pass (with an inserted leading zero) gives the sequence

$$(0201102002)_F \rightarrow (1002102002)_F \rightarrow (1011002002)_F$$
$$\rightarrow (1011010012)_F \rightarrow (1011010101)_F.$$

The second pass gives the sequence $(01011010101)_F \rightarrow (10000010101)_F$, as before.

4.2 Subtraction

As with addition, subtraction can be done in a variety of ways. The simplest is a form of reallocation, as currently taught in most American elementary schools. Digit by digit, we have $0 - 0 = 0$, $1 - 0 = 1$, and $1 - 1 = 0$. In the $0 - 1$ case, we go to the left in the digits of the first number, and find the first 1. Applying the pair rule in reverse then replaces the triple of digits 100 by 011. Repeat this with the right most of the new ones until there is a one in the first number available for canceling. The rightmost digit is a special case, where we replace $(10)_Z$ by $(02)_Z$ before subtraction if necessary.

After using this technique, the number may not be in Zeckendorf form, but it will certainly not contain any twos. Therefore, a single pass of the new approach described for addition will reduce it to Zeckendorf form. Thus, at worst, subtraction can be done in essentially three passes. In the worst case the reallocation steps require moving up and down the entire first number (two passes at worst), then one pass to eliminate pairs of ones. For example, consider $(101000010)_Z - (010000101)_Z$. Matching digits by reallocation requires three applications of the pair rule, replacing the problem by $(011101102)_Z - (010000101)_Z = (1101001)_Z$. Eliminating pairs gives $(10001001)_Z$, the final answer.

It is worth mentioning that Fenwick [6] initially recommended this reallocation approach, then went on to a much more complicated complement approach. Tee [16] also had a slow $O(n^3)$ algorithm. Ahlbach et al. [1] just subtracted digits and added an additional pass to eliminate negative digits. They then used their addition algorithm, resulting in four passes in total. The approach here is the most efficient.

4.3 Multiplication

The literature records four distinct methods for performing multiplication. In the same way that traditional multiplication is performed digit by digit, Freitag and Phillips [8] multiplied numbers by adding the products of Fibonacci numbers that appear in the Zeckendorf representations of the numbers. They prove the odd and even rules

$$f_m f_{2i} = \sum_{j=0}^{i-1} f_{m+2i-2-4j}, \quad \text{and} \quad f_m f_{2i+1} = f_{m-2i} + \sum_{j=0}^{i-1} f_{m+2i-1-4j},$$

where $m > 2i$ and $2i + 1$, respectively. They recommend converting back to Zeckendorf form after each addition to avoid arbitrarily large digits.

Tee [16] suggested using Russian Peasant multiplication, which can be written algebraically as: if y is even, then $xy = (2x)(y/2)$, else $xy = x + x(y - 1) = x + (2x)((y - 1)/2)$. Doubling a number in Zeckendorf form is easy: just replace every one by two, and return to Zeckendorf form. This requires three passes using the new technique. Halving is just as easy using the reversed pair rule, $(\cdots (-1)(+1)(+1) \cdots)$. From left to right, apply the reversed pair rule to any one or three (threes can accumulate from multiple additions). Twos are initially ignored, and after the pass all digits will be zero or two, apart from possibly the last pair. To halve, replace twos by ones, and the last pair of digits identify whether the number was initially odd. Looking at the special cases of halving the last two digits, with T for an odd number and F for not odd, we have

$$(00) \rightarrow (00)F, \qquad (01) \rightarrow (00)T, \qquad (10) \rightarrow (01)F,$$

$$(11) \rightarrow (01)T, \qquad (12) \rightarrow (10)F, \qquad (20) \rightarrow ((10)F,$$

$$(21) \rightarrow (10)T, \qquad (22) \rightarrow (11)F, \qquad (31) \rightarrow (11)T.$$

One more pass will be required to eliminate pairs of ones. For example, with 45, we have $(10010100)_F \rightarrow (01110100)_F \rightarrow (00220100)_F \rightarrow (00220011)_F$. Halving and applying the (11) last pair rule, we get $(110001)_F$, with "true" meaning odd. One last pass replaces this by $(1000001)_F = 22$, which is half of 45, which is odd.

Fenwick [6] suggested a variant of Egyptian multiplication. This version doesn't use doubling and instead adds the previous two numbers. Unfortunately, this technique is less effective than Russian Peasant, because there are more additions than doubles.

A final technique for multiplication is motivated by John Napier's method of addition and multiplication in binary using a checkerboard, as described in Gardner [9]. To begin, label the rows and columns with the Fibonacci numbers used in Zeckendorf form. Then, since each number is represented by a sum of Fibonacci numbers, by the distributive rule their product can be laid out as counters, where each column is associated with the first number, and each row with the second. For example, consider 25×18, or $(1000101)_Z \times (101000)_Z$. Figure 17.2 shows how this would be laid out initially. I have circled the relevant Fibonacci numbers used to represent the first and second numbers.

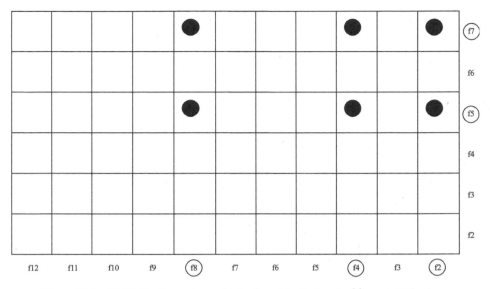

Figure 17.2. Multiplication using a checkerboard in Zeckendorf form, initial setup.

We now want to return the number to Zeckendorf form, where we can now work down columns as well as across rows. One systematic approach is to use the pair rule in reverse in the vertical direction to remove every counter in the top row, then use the addition approach to return the rows that have changed back to Zeckendorf form. We repeat this process until a single row remains at the bottom. As with addition, some care needs to be taken with the final step. Figure 17.3 shows the intermediate steps for 25 × 18. While it may not look terribly obvious as a figure, physically moving counters around a checkerboard and applying the pair and two rules is an easy process to follow. It is also very easy to implement on a computer using a two-dimensional array. There is a similarity here to the Freitag and Phillips [8] approach, but it is not immediately obvious whether the two-dimensional array cuts down the amount of work required to find the solution.

Note that Ahlbach et al. [1] avoid multiplication by converting to binary, multiplying in binary, and then converting back to Zeckendorf form.

5 Conclusion

We have seen how Zeckendorf notation is an excellent way of representing a stream of natural numbers, particularly when either the maximum of the stream is not known beforehand, or smaller numbers are more likely than larger ones. The technique is particularly useful when representing the

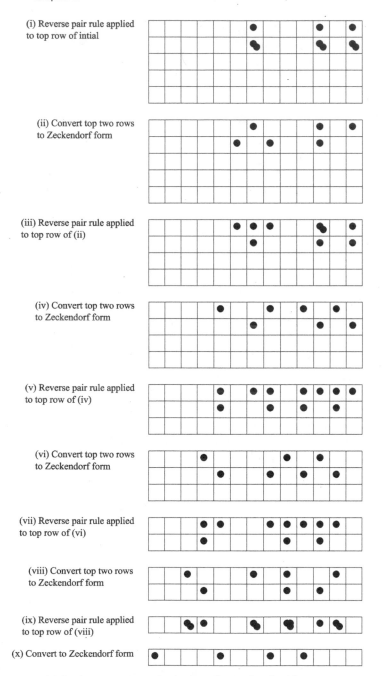

(i) Reverse pair rule applied
to top row of intial

(ii) Convert top two rows
to Zeckendorf form

(iii) Reverse pair rule applied
to top row of (ii)

(iv) Convert top two rows
to Zeckendorf form

(v) Reverse pair rule applied
to top row of (iv)

(vi) Convert top two rows
to Zeckendorf form

(vii) Reverse pair rule applied
to top row of (vi)

(viii) Convert top two rows
to Zeckendorf form

(ix) Reverse pair rule applied
to top row of (viii)

(x) Convert to Zeckendorf form

Figure 17.3. Multiplication using a checkerboard in Zeckendorf form, successive steps.

sequence of partial quotients that make up the continued fraction representation of an arbitrary irrational. While it is possible to generalize to sequences where each is the sum of more than two previous numbers, this turns out to be of limited utility.

We have also seen how addition and subtraction of numbers in Zeckendorf form is a straightforward task and have shown a new approach to returning a sequence to Zeckendorf form that is more efficient. We have also seen four different ways of multiplying, and that using a checkerboard as a calculation tool has a certain elegance.

There are a number of future directions to consider. Which of these multiplication algorithms is most efficient? Does this depend on the magnitude of the numbers? Some of the authors already cited have suggested integer division (quotient and remainder) algorithms. How do they compare, particularly with the new efficient addition algorithm presented here? In addition, Bunder [3] showed how all integers can be represented by Fibonacci numbers with negative coefficient, and Anderson and Bicknell-Johnson [2] showed how vectors built on k-bonacci numbers can be used to represent points in $\mathbb{Z}^{k-1}$. Can we do arithmetic on these in the same way?

References

[1] C. Ahlbach, J. Usatine, C. Frougny, and N. Pippenger. Efficient algorithms for Zeckendorf arithmetic. *Fibonacci Quart.* **51** no. 3 (2013) 249–255.

[2] P. G. Anderson and M. Bicknell-Johnson. Multidimensional Zeckendorf representations. *Fibonacci Quart.* **49** no. 1 (2011) 4–9.

[3] M. W. Bunder. Zeckendorf representations using negative Fibonacci numbers. *Fibonacci Quart.* **30** no. 2 (1992) 111–115.

[4] R. A. Dunlap, *The Golden Ratio and Fibonacci Numbers.* World Scientific, Singapore, 1998.

[5] M. Feinberg. Fibonacci-tribonacci. *Fibonacci Quart.* **1** no. 3 (1963) 71–74.

[6] P. Fenwick. Zeckendorf integer arithmetic. *Fibonacci Quart.* **41** no. 5 (2003) 405–413.

[7] A. S. Fraenkel, Systems of numeration. *Am. Math. Monthly* **92** no. 2 (1985) 105–114.

[8] H. T. Freitag and G. M. Phillips. Elements of Zeckendorf arithmetic, in G. E. Bergum, A. N. Philippou and A. F. Horadam, editors, *Applications of Fibonacci Numbers,* Volume 7, p. 129. Kluwer, Dordrecht, 1998.

[9] M. Gardner. *Knotted Doughnuts and Other Mathematical Entertainments.* W. H. Freeman and Company, New York, 1986.

[10] G. H. Hardy and E. M. Wright. *An Introduction to the Theory of Numbers,* fifth edition. Oxford University Press, London, 1979.

[11] A. Ya. Khinchin. *Continued Fractions.* Dover, Mineola, NY, 1997 (originally published by University of Chicago Press, 1964).

[12] C. G. Lekkerkerker. Voorstelling van natuurlijke getallen door een som van getallen van Fibonacci. *Simon Stevin* **29** (1952) 190–195.

[13] M. Livio. *The Golden Ratio: The Story of φ, the World's Most Astonishing Number.* Broadway Books, New York, 2003.

[14] G. Lochs. Vergleich der Genauigkeit von Dezimalbruch und Kettenbruch. *Abh. Hamburg Univ. Math. Sem.* **27** (1964) 142–144.

[15] C. D. Olds. *Continued Fractions.* Random House, New York, 1963.

[16] G. J. Tee. Russian peasant multiplication and Egyptian division in Zeckendorf arithmetic. *Austral. Math. Soc. Gaz.* **30** no. 5 (2003) 267–276.

[17] S. Vajda. *Fibonacci and Lucas Numbers, and the Golden Section: Theory and Applications.* Dover, Mineola, NY, 2007.

[18] E. W. Weisstein. Lochs' theorem. MathWorld—A Wolfram Web Resource. http://mathworld.wolfram.com/LochsTheorem.html (accessed April 2015)

[19] E. Zeckendorf. Représentation des nombres naturels par une somme de nombres de Fibonacci ou de nombres de Lucas. *Bull. Soc. Roy. Sci. Liége* **41** (1972) 179–182.

About the Editors

Jennifer Beineke is a professor of mathematics at Western New England University, Springfield, Massachusetts. She earned undergraduate degrees in mathematics and French from Purdue University, West La Fayette, Indiana, and obtained her PhD from University of California, Los Angeles in 1997. She held a visiting position at Trinity College, Hartford, Connecticut, where she received the Arthur H. Hughes Award for Outstanding Teaching Achievement. Her research in the area of analytic number theory has most recently focused on moments of the Riemann zeta function. She enjoys sharing her love of mathematics, especially number theory and recreational mathematics, with others, usually traveling to math conferences with some combination of her husband, parents, and three children.

Jason Rosenhouse is a professor of mathematics at James Madison University, Harrisonburg, Virginia, specializing in algebraic graph theory. He received his PhD from Dartmouth College, Hanover, New Hampshire, in 2000 and has previously taught at Kansas State University, Manhattan. He is the author of the books *The Monty Hall Problem: The Remarkable Story of Math's Most Contentious Brainteaser* and *Among the Creationists: Dispatches from the Anti-Evolutionist Front Line.* With Laura Taalman, he is the coauthor of *Taking Sudoku Seriously: The Math Behind the World's Most Popular Pencil Puzzle.* All three books were published by Oxford University Press. He is also the editor of *Four Lives: A Celebration of Raymond Smullyan,* published by Dover. When not doing math he enjoys chess, cooking, and reading locked-room mysteries.

Max A. Alekseyev is an associate professor at the Computational Biology Institute and the Department of Mathematics at the George Washington University Washington, DC. He received an MS in mathematics in 1999 from the N. I. Lobachevsky State University of Nizhni, Novgorod, Russia, and a PhD in 2007 in computer science from the University of California, San Diego. His research interests include computational molecular biology, graph theory, combinatorics, and discrete algorithms. He received a National Science Foundation CAREER award in 2013. He is an associate editor of the *Frontiers in Bioinformatics and Computational Biology* journal and Editor-in-Chief of the On-Line Encyclopedia of Integer Sequences®.

Julie Beier is a representation theorist turned algebraic combinatorialist turned flexagoner. She also has hidden faces in graph theory and mathematics education. She strives to be a truly liberal arts teacher, diving into issues as diverse as social justice, contemplative education, and political philosophy with her students. She works at Earlham College, Richmond, Indiana.

Lowell Beineke is Schrey Professor of Mathematics at Indiana University-Purdue University Fort Wayne, where he has worked since receiving his PhD from the University of Michigan, Ann Arbor, in 1965. His graph theory interests include topological graph theory, line graphs, tournaments, and decompositions. He has published more than a hundred papers and served a term as editor of the *College Mathematics Journal.* He and Robin Wilson have edited nine books on topics in graph theory, the most recent being *Topics in Chromatic Graph Theory.* Honors include being entered into Purdue University's *Book of Great Teachers* and a Certificate of Meritorious Service from the Mathematical Association of America. He enjoys sharing math (and grandchild activities) with his mathematician daughter and co-author, and his statistician son.

Toby Berger is a professor in the Electrical and Computer Engineering Department of the University of Virginia, Charlottesville. Since the late 1990s, his principal research interest has been the application of information theory to problems in neuroscience. He has received a Guggenheim Fellowship, the American Society For Engineering Education Terman Award and the Institute of Electrical and Electronics Engineers (IEEE) Shannon, Kirchmayer, and Wyner Awards and Hamming Medal. He is a Life Fellow of the IEEE, a member of the National Academy of Engineering, and an avid amateur blues harmonica player.

Robert Bosch is a professor of mathematics at Oberlin College, Ohio, and an award-winning writer and artist. He specializes in optimization, the branch of mathematics concerned with optimal performance. Since 2001, Bosch has thrown himself into devising and refining methods for using optimization to create visual artwork. He has had pieces commissioned by Colorado College, Western Washington University, Occidental College, Spelman College, and the organizing committees of several academic conferences. He operates www.dominoartwork.com, from which one can download free plans for several of his domino mosaics. He is hard at work on a book on optimization and the visual arts.

Leigh Marie Braswell is a graduate of Phillips Exeter Academy, Exeter, New Hampshire, and a freshman at the Massachusetts Institute of Technology, Cambridge. She is especially interested in combinatorics and mathematical puzzles. She also enjoys computer programming and dancing.

Neil Calkin has studied at Trinity College, Cambridge, and at the University of Waterloo, Ontario, and has attempted to inspire students with magic since his friend Colm tricked him the first time. He is proud of his Erdős number 1, and is hoping for an invitation to appear in a movie with Kevin Bacon. He is a professor of mathematics at Clemson University, South Carolina.

Maureen T. Carroll was a fellow in Project NExT and a participant in the Institute in the History of Mathematics and Its Use in Teaching, both programs sponsored by the Mathematical Association of America (MAA). She received the James P. Crawford Award for Distinguished College or University Teaching from the MAA's Eastern Pennsylvania and Delaware section in 2013. She and her co-author Steven Dougherty met during their first week of graduate school. They were awarded the MAA's Hasse Prize in 2005 for their article "Tic-tac-toe on a Finite Plane in *Mathematics Magazine.*

Tim Chartier teaches mathematics and computer science at Davidson College, North Carolina. He is the author of *Math Bytes: Google Bombs, Chocolate-Covered Pi, and Other Cool Bits in Computing.* Tim is a member and past chairperson of the Advisory Council for the National Museum of Mathematics and was named the first Math Ambassador of the Mathematical Association of America. He is a recipient of a national teaching award and an Alfred P. Sloan Research Fellowship. He fields mathematical questions for ESPN's *Sport Science* program and has served as a resource for the *CBS Evening News*, National Public Radio, *The New York Times*, and other major news outlets.

Steven T. Dougherty received his PhD from Lehigh University, Bethlehem, Pennsylvania. He has written more than seventy-one papers in coding theory, number theory, combinatorics, and algebra with forty-two different co-authors and has lectured in twelve countries. He is currently a professor of mathematics at the University of Scranton, Pennsylvania, where he teaches with his co-author Maureen Carroll.

Gary Gordon is a professor of mathematics at Lafayette College, Easton, Pennsylvania. He is interested in matroids, combinatorics, and finite geometry, and with Jennifer McNulty, he co-authored the text *Matroids: A Geometric Introduction*. He ran Lafayette's Research Experience for Undergraduates program from 2000 to 2010 and has published extensively with undergraduates. He is currently the Problem Section editor for *Math Horizons*. He has won awards for his teaching and his research from Lafayette, and is proud to have served as a Posse mentor. He is the slowest SET® player in the family.

Tanya Khovanova is a lecturer at the Massachusetts Institute of Technology, Cambridge, and a freelance mathematician. She received her PhD in mathematics from the Moscow State University in 1988. At that time, her research interests were in representation theory, integrable systems, superstring theory, and quantum groups. Her research was interrupted by a period of employment in industry, where she became interested in algorithms, complexity theory, cryptography, and networks. Her current interests lie in recreational mathematics, including puzzles, magic tricks, combinatorics, number theory, geometry, and probability theory. Her website is located at tanyakhovanova.com, her math blog at blog.tanyakhovanova.com, and her Number Gossip website at numbergossip.com.

Dominic Lanphier is an associate professor at Western Kentucky University, Bowling Green. He obtained his PhD from the University of Minnesota, Minneapolis, and his undergraduate degree from the University of Michigan, Ann Arbor. He has held postdoctoral positions at Oklahoma State University, Stillwater, and Kansas State University, Manhattan. His research interests lie in number theory, where he studies automorphic forms and L-functions, and in discrete mathematics.

Anany Levitin graduated from the Moscow State University with an MS in mathematics. He holds a PhD in mathematics from the Hebrew University of Jerusalem and an MS degree in computer science from the University of Kentucky, Lexington. He currently is a professor of computing sciences at Villanova University, Pennsylvania. From 1990 to 1995, he also worked as a consultant at AT&T Bell Laboratories. In addition to several dozen papers, he authored two books: *Introduction to the Design and Analysis of Algorithms*, which has been translated into five foreign languages, and, jointly with Maria Levitin, *Algorithmic Puzzles*, translated into Chinese and Japanese.

Stephen K. Lucas received his bachelor of mathematics from the University of Wollongong, New South Wales, in 1989 and his PhD from the University of Sydney in 1994. In 2002 he received the Michell Medal for Outstanding New Researchers from Australian and New Zealand Industrial and Applied Mathematics, Australia. He is currently a professor at James Madison University, following a postdoc at Harvard and a faculty position at the University of South Australia. His research interests span a wide range of topics in applied and pure mathematics, usually with a numerical bent. He "fell into Fibonacci" while studying the history of number representations.

Elizabeth McMahon is a professor of mathematics at Lafayette College, Easton, Pennsylvania. She has worked in several areas of mathematics, most recently the affine geometry of the card game SET®. She collaborates frequently with Gary Gordon and also wrote an article about SET® for PRIMUS with her two daughters. Her whole family is writing a book about the game, to be published by Princeton University Press. She has worked to increase the success of underrepresented groups in mathematics and has won several teaching awards at Lafayette, as well as the Mathematical Association of America's EPaDel section James P. Crawford Award for Distinguished Teaching of Mathematics.

John K. McSweeney is an assistant professor of mathematics at Rose-Hulman Institute of Technology, Terre Haute, Indiana. He grew up in Montreal, Canada, where he graduated from McGill University, and went on to do his PhD in probability theory at Ohio State University, Columbus. As a child of English teachers, he has always enjoyed words and languages, and is constantly looking to apply his love of mathematics to real-world settings. His chapter on the mathematics of crossword puzzles is a perfect marriage of those interests, although it was inspired by more "serious" work on the spread of epidemics that he developed during a visit to the Statistical and Applied Mathematical Sciences Institute, Research Triangle Park, North Carolina.

David Molnar received his PhD from the University of Connecticut, Mansfield, in 2010. His mathematical interests include number theory, dynamical systems, graph theory, and, evidently, games. Keeping in mind his own undergraduate experience, including a term with the Budapest Semesters in Mathematics, he strives to broaden students' awareness of what constitutes mathematics. One of his outlets for doing so is competitions; since 2009, he has been involved with the New Jersey Undergraduate Mathematics Competition. He is currently a lecturer at Rutgers University New Brunswick, New Jersey.

Colm Mulcahy recently completed ten years of "Card Colms" at MAA.org, bi-monthly columns exploring mathemagic with playing cards, very much inspired by the spirit of Martin Gardner. He recently published the book *Mathematical Card Magic: Fifty-Two New Effects*, of mostly original principles and creations. He is a professor of mathematics at Spelman College, Atlanta, Georgia.

Michael Rowan is a PhD student at Harvard University, Cambridge, Massachusetts, studying high-energy physics. Recently, as a Teaching Fellow at Harvard summer school, he has gained exposure teaching physics to high school and college students. His undergraduate studies were in physics and mathematics at Oberlin College, Ohio. He has worked on research projects aimed at understanding the structure of atoms through high-precision measurement, as well as projects in mathematical art related to both maze design and fractal art. In his spare time, he likes to play the piano and explore the Boston area.

Derek Smith is an associate professor of mathematics at Lafayette College, Easton, Pennsylvania. He is the co-author with John Horton Conway of *On Quaternions and Octonions: Their Geometry, Arithmetic, and Symmetry*, and is a former editor of the "The Playground," the problem section of *Math Horizons*.

Laura Taalman is a professor in the Department of Mathematics and Statistics at James Madison University, where she joined the faculty after graduate work at Duke University and undergraduate work at the University of Chicago. This year, she is serving as the mathematician-in-residence at MoMath, the National Museum of Mathematics. Her mathematical research interests include singular algebraic geometry, knot theory, and the mathematics of games and puzzles. She is also a co-author with Peter Kohn of the recent *Calculus* textbook and seven books on Sudoku and mathematics. In 2013 she was a recipient of the Outstanding Faculty Award from the State Council of Higher Education for Virginia.

Robert W. Vallin earned his PhD in mathematics from North Carolina State University, Raleigh, in 1991. In addition to publishing in classical real analysis, his original research area, his technical papers cover topology and fractal geometry. He is also the author of pedagogical and expository publications. Always interested in bringing novel ideas to the classroom, for the past five years his interest in recreational math has expanded. His forays into this topic include card magic and KenKen puzzles, and he has guided undergraduate research on juggling, card tricks, and games. He is currently at Lamar University, Beaumont, Texas.

Peter Winkler is William Morrill Professor of Mathematics and Computer Science at Dartmouth College, Hanover, New Hampshire. He is the author of about 150 research papers and holds a dozen patents in computing, cryptology, holography, optical networking, and marine navigation. His research is primarily in combinatorics, probability, and the theory of computing, with forays into statistical physics. He has also written two collections of mathematical puzzles, a book on cryptology in the game of bridge, and a portfolio of compositions for ragtime piano. And he is a big fan of MoMath!

Carolyn Yackel is a commutative algebraist turned mathematical fiber artist turned recreational mathematician, who also participates in mathematics education research. With all of those faces, one can see why she is fascinated by flexagons. She is an innovative educator who elegantly guides students to discover mathematics, and works at Mercer University, Macon, Georgia.

Index